Confined Magnon Modes and Anisotropic Exchange Interaction in Ultrathin Co Films

Dissertation

zur Erlangung des Doktorgrades der Naturwissenschaften
(Dr. rer. nat.)

der

Naturwissenschaftlichen Fakultät II
Chemie, Physik und Mathematik

der Martin-Luther-Universität
Halle-Wittenberg

vorgelegt von

Frau **Ying-Jiun Chen**
geb. am 31.10.1984 in Taitung, Taiwan

Gutachter:

1. Prof. Dr. Jürgen Kirschner

2. Prof. Dr. Georg Woltersdorf

3. Prof. Dr. Wulf Wulfhekel

Tag der Verteidigung: 01.12.2016

Bibliographic information published by the Deutsche Nationalbibliothek

The Deutsche Nationalbibliothek lists this publication in the Deutsche
Nationalbibliografie; detailed bibliographic data are available
on the Internet at http://dnb.d-nb.de .

ISBN 978-3-8325-4470-6

Logos Verlag Berlin GmbH
Comeniushof, Gubener Str. 47,
10243 Berlin
Tel.: +49 (0)30 42 85 10 90
Fax: +49 (0)30 42 85 10 92
INTERNET: http://www.logos-verlag.de

Contents

Chapter 1

Introduction

The coupling of spins by exchange interaction lies at the heart of magnetism. An interplay of different terms in the magnetic free energy results in a large variety of magnetic structures such as simple ferromagnetic and antiferromagnetic structures, spiral arrangement, helical order, noncollinear magnetic phase, and frustrated systems [1–3]. One of the most important terms in the free energy is the magnetic exchange energy. A qualitative and quantitative understanding of the magnetic exchange interaction is imperative for designing novel materials with desired properties and for downsizing magnetic systems to the nanoscale. Investigation of elementary magnetic excitations and a quantitative determination of the effective exchange parameters in low-dimensional ferromagnets are the core subjects of this thesis.

In a ferromagnet, the strong exchange interaction between neighboring spins results in a parallel alignment of spins, forming a fully ordered magnetic ground state. The magnetic excitations of such an ordered system can be rigorously regarded as a perturbation (or excitation) of the respective ground state. Collective spin-wave excitations can be described by bosonic quasiparticles. Their corresponding quanta are named magnons. The energies of magnons span several orders of magnitude in the range from μeV to a few eV. Their wavelengths (frequencies) vary from microns (GHz) to nm (THz), governed by the weak long-range magnetic dipole-dipole interaction and the strong short-range exchange interaction, respectively. Exploring the characteristics of magnons is essential to understand numerous physical properties in magnets, including the specific heat, electrical and thermal resistivity, etc. Moreover, it is generally believed that the coupling between electrons and magnetic excitations has bearing on the pairing mechanism of electrons in unconventional superconductors [4–7].

Understanding of exchange-dominated magnons is of great importance for both fundamental science and modern applications, e.g., in the new field of magnonics. Since magnons allow the transfer of spin angular momentum without charge transport, magnon based devices for data processing and computation might be realized free of charge-related dissipation [8, 9]. As the exchange-dominated magnons with their nanometer wavelengths reach into the THz frequency range, this opens a new avenue to meet the modern computing demands: downsizing, ultrafast, and energy-efficient data processing at room temperature [8].

Exploring the coupling of collective and particle-hole excitations in itinerant magnets is a focused topic in magnetism. For itinerant magnets, the exchange magnons are strongly influenced by the electron–hole pair excitation. They tend to be damped by coupling to these excitations. The attenuation of the exchange magnons in itinerant magnets is usually accounted by the hybridization with the continuum of noncoherent particle-hole Stoner excitations, referred to as Landau damping. In contrast to "localized" bulk magnets, in which the number of magnon modes is restricted by the number of atoms per unit cell, the additional degree of freedom in itinerant magnets may lead to the intriguing "disappearance" or "appearance" of magnon modes. For instance, it has been found that the magnon decay into a Stoner continuum gives rise to vanishing magnon branches at certain points in the Brillouin zone [10–13]. On the other hand, the unexpected presence of the "optical" magnon mode in bulk Ni is still an unsettled question that can not be explained by a Heisenberg model of local moments [14–16], and thus calling for a refined description of magnons in an itinerant system.

The study of many-body correlation effects in itinerant magnets is one of central challenges in condensed matter physics. Notably, the late $3d$ transition metals, Fe, Co and Ni, have been recognized recently as important facets of "strongly" correlated itinerant ferromagnets. It has been experimentally demonstrated that many-body correlation effects in $3d$ transition metals give rise to characteristic anomalies in quasiparticle renormalization [17–19]. Moreover, the renormalized electronic quasiparticle states, arising from the coupling to bosonic excitations (phonons and magnons), have also been observed in photoemission spectra [20, 21]. Although many aspects of many-body correlation effects on photoemission quasiparticle spectra have been demonstrated, some crucial questions, regarding whether or how the nature of elementary magnetic excitations and the exchange interaction in itinerant magnets are influenced by the many-body correlation effects, still remain unanswered and have not been verified experimentally.

Magnon confinement in ultrathin films

One of today's challenges in magnonics is to create, identify, and manipulate the exchange magnons. For this purpose, an ideal candidate is an ultrathin magnetic film. The presence of the vertical confinement in the film eliminates the formation of long-wavelength magnons, dominated by dipole interaction, in the surface normal direction. Then, only confined magnon modes governed by exchange interaction are allowed to be formed. The number of confined exchange magnon modes can be adjusted precisely according to the well-defined film thickness in terms of the number of atomic layers. The confinement in ultrathin magnetic films therefore opens up a possibility to engineer discretely different magnon modes in an ultrathin film.

For ultrathin films on substrates, the atomic bonding environment for atoms situated at surface, interior, and interface layers is substantially different from each other, due to the presence of the distinct vacuum/film and substrate/film interfaces. These differences result in distinguished physical properties. For instance, the lowered coordination number at the film surface leads to an enhanced magnetic moment [22, 23]. At the interface, however, this effect may be suppressed by strong electronic hybridization between the film and substrate [24, 25]. Therefore, one would expect at least three different types of magnons (surface, interior, and interface magnons) with own characteristics (e.g., energy, velocity, amplitude, and lifetime) being present in the system.

More importantly, the magnetic exchange interaction in a magnetic film can be altered by modifying the lattice strain, atomic structure, film thickness, the choice of substrate, or inserting different magnetic elements. These possible modifications offer a versatile platform for tailoring their functionality.

Since a few atomic-layers-thick magnetic films provide unprecedented access to engineer the confined magnon modes in a precisely controlled way, a complete characterization of the distinct confined magnon modes can be achieved. By probing the confined magnon modes, the influence of many-body correlation effects on the properties of individual magnon modes and the strength of exchange interaction in low-dimensional itinerant magnets can be experimentally addressed.

Spin resolved inelastic electron scattering

Magnetic excitations in solids are mainly investigated by inelastic-scattering experiments. Particles such as neutrons, photons, and electrons can be scattered inelastically as they interact with elementary excitations (e.g. phonons, plasmons and magnons) in a material. During the inelastic scattering process, the energy and momentum of the excited quasiparticles can then be obtained as the energy loss and the momentum transfer of the probing particles, respectively. In addition, the spectral linewidths give information about the excitation lifetimes.

However, probing elementary excitations in low-dimensional systems is challenging by means of both neutron and photon scattering techniques, since the interaction between these probing particles and the solid is rather weak. Particularly, the detection of magnetic excitations, up to the surface Brillouin zone boundary, in ultrathin itinerant magnets is beyond the capabilities of inelastic neutron and photon (e.g., Brillouin light scattering and resonant inelastic x-ray scattering) scattering experiments. A key physical quantity, the interatomic exchange interaction, in low-dimensional magnets is therefore hardly detectable. Inelastic electron scattering experiments provide a unique way of probing magnetic excitations in low-dimensional solids. Besides neutrons and photons as a probe, the use of electrons has following prominent advantages:

- The strong interaction between incident electrons and solids gives rise to higher scattering probabilities. This results in a high surface sensitivity (rather short penetration length) and a large inelastic scattering cross section.

- By using spin-polarized electrons as a probe, signals contributed from magnetic excitations in magnets can be distinguished efficiently from other spin-independent excitations.

- Electrostatic and magnetic lens systems provide a straightforward means to manipulate and to analyze electron beams. The mandatory high energy and momentum resolution can therefore be routinely achieved.

These advantages make electrons an ideal probe to investigate elementary excitations and to directly map their dispersion relations over the whole surface Brillouin zone in low-dimensional systems.

Major advances have been made recently via inelastic electron scattering in accessing localized, collective, and discrete excitations of itinerant nanomagnets [26–30]. The progress has been triggered by the development of spin-polarized electron energy-loss spectroscopy (SPEELS). The state-of-the-art spin-polarized electron energy-loss spectroscopy permits highly spin-, energy-, and momentum-resolved inelastic scattering measurements over the whole surface Brillouin zone [29,31]. This technique is also widely applicable to other elementary excitations in solids, such as phonons, plasmons, electron-hole pairs, and crystal-field excitations, that holds promise to study many fascinating many-body phenomena and the underlying fundamental mechanisms.

Aims and scope of this work

The aim of this study is the experimental investigation of elementary magnetic excitations in ultrathin Co films on different substrates. Measurements are performed with precisely known film thicknesses in terms of the number of atomic layers. The main idea of the work is to explore the nature of confined magnon modes and the corresponding magnon dispersion relation by utilizing SPEELS.

Ultrathin Co films are chosen in this work, since Cobalt has the highest ferromagnetic Curie temperature ($T_C = 1388$ K) and large magnetic anisotropy among the $3d$

transition metals. Moreover, Cobalt serves as an example of a "strongly" correlated itinerant ferromagnet and has its electronic structure dominated by exchange-split d bands in the vicinity of the Fermi level. The pronounced spin-dependent many-body correlations lead to spin-dependent renormalization of electronic quasiparticle states and the quenching of majority-channel quasiparticle excitations [18,32]. Cobalt films studied here offer a clear-cut low-dimensional itinerant ferromagnetic system to clarify the many-body correlations effects on the nature of the confined exchange magnon modes and the effective interatomic exchange interactions, as will be discussed in this work.

By comparing the experimental results to the results of first-principles calculations, the strength of layer-dependent exchange interaction in ultrathin Co films on different substrates could be quantitatively determined. The distinct magnon modes localized at surface, interior, and interface layers of an ultrathin Co film were characterized. Furthermore, the influence of the substrates, magnetic anisotropy, epitaxial strain, atomic structures, film thickness, and substrate surface reconstructions on the properties of distinct magnon modes are investigated.

The important questions, which will be addressed in this work are: (i) How does the magnon dispersion relation change when different confined magnon modes are presented in the system? (ii) How does the dispersion relation of distinct magnon modes depend on the relative strength of the intra- and inter-layer exchange interaction? (iii) How the ground-state electronic structures and the substrate-derived interface states affect the different magnon modes? (iv) How do many-body effects influence the characteristics of the different magnon modes, the corresponding dispersion relation, and the layer-dependent exchange parameters? All of these questions are the subject of the current investigation.

The structure of this thesis is organized as follows. In Chapter 2, the quasi-classical and quantum mechanical descriptions of a magnon in localized moment systems, and the theoretical background for the interpretation of confined magnon modes in ultrathin ferromagnetic films are provided. In Chapter 3, the basic concept and the experimental setup of spin-polarized inelastic electron scattering and the principle of SPEELS are introduced. In Chapter 4, the experimental results of sample preparation and characterization as well as SPEELS measurements are presented. In the course of this work, a detailed study, regarding the effects of the substrate, magnetic anisotropy, epitaxial strain, atomic structures, film thickness, and substrate surface reconstructions on the properties of distinct magnon modes is demonstrated. In Chapter 5, the discussion of the results and a comparison to the results of *ab initio* calculations are provided. The experimental findings are also compared to results of different sample systems obtained from other studies. In Chapter 6, the major aspects in this work are briefly summarized and an outlook for future studies is also provided.

Chapter 2

Background

2.1 Magnons in localized moment systems

Although the exchange integral (J) is of quantum-mechanical origin with no classical analogue, the energy of magnons (spin waves) as well as their corresponding precessional modes can be derived from the Heisenberg model in a way that treats the spin as a classical vector. The results are equivalent to those obtained by the quantum mechanical description. The main difference between this approach and the fully quantum mechanical description is, that in the latter case the three "vector" components of the spin are not commensurate. In this section, we will discuss how magnons arise from the magnetic ground state and are described by means of quasi-classical and quantum-mechanical approaches from the Heisenberg Hamiltonian. For simplicity, we consider only the case of localized magnetic moments on each lattice site associated with the spin moment (negligible orbital contributions).

2.1.1 The quasi-classical counterpart of a magnon

In a quantum mechanical view, the quantized vibrational excitations of the crystal lattice in solids are called phonons (particle-like properties). Their corresponding normal modes of vibration (wave-like phenomena) can be describe by classical mechanics. Similarly, magnons can also be described by using a model of a precessing spin. In the quantum mechanical description, this corresponds to a one-magnon eigenstate, as we will demonstrate in Sec. 2.1.2.

For the Heisenberg ferromagnets with $S = \frac{1}{2}$, the lowest-energy excited state can be visualized by a single flipped spin, distributed over the ensemble of spins. In the classical description, we assume that the spin operator is substituted by the length of the classical vector of the spin, $\sqrt{S(S+1)} \simeq S$. The values of S_i^z are limited to be $-S, -S+1, \ldots, S-1, S$ where S is the spin quantum number. In the limit of a large S, the classical version of magnons can be described by the precession of

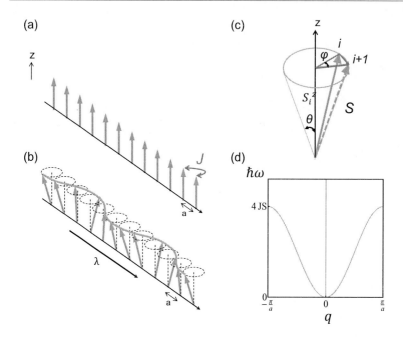

Figure 2.1: Quasi-classical magnons propagate along a direction perpendicular to the z-axis in a linear ferromagnetic chain of lattice constant a. (a) In the ground state all spins are aligned parallel in the z direction. (b) A quasi-classical magnon state at a given wave-vector, $q = \frac{2\pi}{\lambda}$, in a ferromagnetic chain, as an analogue for the eigenstate $|q\rangle$ in Eq. (2.15). (c) By evolving time, all spins precess on a cone. The angle θ refers to the deviation of the spin from the z-axis. The precession angle of the spins varies from site-to-site in the propagation direction. Each spin differs in phase from the prior one by an angle of $\varphi = \mathbf{q} \cdot (\mathbf{r_i} - \mathbf{r_j})$. (d) The magnon dispersion relation within the first Brillouin zone ($q = \pm\frac{\pi}{a}$) at zero external magnetic field ($\mathbf{H_{ext}} = 0$) for a linear ferromagnetic chain.

spin vectors around an axis, which is determined by a small applied magnetic field or the effective anisotropy field. Magnons with a given wave-vector q have a fixed phase relation among the precessing spins, as shown in Fig. 2.1b.

We begin with replacing the spin operator by a form of the angular momentum $\mathbf{L} = \hbar\mathbf{S_i}$. For simplicity, we consider a linear ferromagnetic chain of lattice constant a with conditions of periodical boundaries. The magnetic moment on the site i is $\mathbf{m_i} = g\mu_B\mathbf{S_i}$. In an effective small magnetic field $\mathbf{H_{eff}}$, the torque is equivalent to the rate of change of the angular momentum, $\frac{d\mathbf{L}}{dt} = \mathbf{m_i} \times \mathbf{H_{eff}}$, which is given by

$$\frac{d}{dt}(\hbar\mathbf{S_i}) = \mathbf{m_i} \times \mathbf{H_{eff}} = g\mu_B\mathbf{S_i} \times \mathbf{H_{eff}}. \tag{2.1}$$

$\mathbf{H_{eff}}$ includes the external magnetic field ($\mathbf{H_{ext}}$) and the effective field arising from the exchange interaction of the nearest-neighbor spins and can be expressed as[1]

$$\mathbf{H_{eff}} = \mathbf{H_{ext}} + \frac{2J}{g\mu_B}\sum_j \mathbf{S_j} \qquad (2.2)$$

By substituting Eq. (2.2) into Eq. (2.1), the torque acting on the spin i is

$$\frac{d\mathbf{S_i}}{dt} = \frac{g\mu_B}{\hbar}\mathbf{S_i} \times \mathbf{H_{i(ext)}} + \frac{2J}{\hbar}\sum_j \mathbf{S_i} \times \mathbf{S_j}. \qquad (2.3)$$

Assuming that the magnetic field is applied along the z-axis and a magnon propagates along a direction perpendicular to the z-axis. The spins deviate only little from the z-direction, i.e., $S^z \simeq S$ and S^x, $S^y \ll S$. As a result, three scalar equations of Eq. (2.3) can be expressed by

$$\frac{dS_i^x}{dt} = \frac{g\mu_B}{\hbar}S_i^y H_{ext} + \frac{2J}{\hbar}[S_i^y S_j^z - S_j^y S_i^z], \qquad (2.4)$$

$$\frac{dS_i^y}{dt} = -\frac{g\mu_B}{\hbar}S_i^x H_{ext} - \frac{2J}{\hbar}[S_i^x S_j^z - S_j^x S_i^z], \qquad (2.5)$$

and

$$\frac{dS_i^z}{dt} = 0 \qquad (2.6)$$

The solutions of above equations are

$$S_i^x = S\sin\theta\cos(\mathbf{q}\cdot\mathbf{r_i} - \omega t), \qquad (2.7)$$

$$S_i^y = S\sin\theta\sin(\mathbf{q}\cdot\mathbf{r_i} - \omega t), \qquad (2.8)$$

and

$$S_i^z = S\cos\theta \qquad (2.9)$$

where $\mathbf{r_i}$ is the position of lattice atom i, $\mathbf{q}$ is the wave-vector of magnons, $\hbar\omega$ is the energy of magnons, and the angle θ refers to the deviation of the spin from the z-axis (see Fig. 2.1c). In the first order approximation of θ, the energy of a magnon as a function of the wave-vector is given by

$$\hbar\omega = g\mu_B H_{ext} + 2JS\cos\theta\sum_r [1 - \cos(\mathbf{q}\cdot\mathbf{r})] \qquad (2.10)$$

where $\mathbf{r}=\mathbf{r_i}-\mathbf{r_j}$.

---

[1]The Hamiltonian of a magnetic moment in the presence of an effective field $\mathbf{H_{eff}}$ is given by $-\mathbf{m_i}\cdot\mathbf{H_{eff}}$. Summed up over all lattice sites, one obtains the same equation as Eq. (2.11). The factor of two in the exchange term is due to summing twice over each pair of spins.

In the classical description of magnons, one can easily see that all spins deviate from their equilibrium orientation. The precession angle of the spins varies from site-to-site in the direction of magnon propagation. Each spin differs in phase from the prior one by a certain angle, $\varphi = \mathbf{q} \cdot (\mathbf{r_i} - \mathbf{r_j})$, so that they form a wave traveling through the lattice (Fig. 2.1b). As the phase angle φ gets larger, more energy is required against the exchange interaction of neighboring spins. As a result, the magnon dispersion relation (the energy versus a wave-vector) over the Brillouin zone is obtained (see Fig. 2.1d).

2.1.2 A magnon in quantum mechanics

Here, we introduce a magnon descried by quantum mechanical operators. The general form of the isotropic Heisenberg Hamiltonian including the Zeeman term is given by

$$\mathcal{H} = -J \sum_{i,j} S_i \cdot S_j - H \sum_i S_i^z \tag{2.11}$$

where i and j denote the lattice sites. $2J = 2J(|\mathbf{r_i} - \mathbf{r_j}|)$ is the exchange constant between spins at i and j, depending on the the the distance between spins. H in the second term is represented by $H = g\mu_B H_{ext}$ where g is the Landé factor, μ_B is the Bohr magneton. H_{ext} is the external magnetic field assumed to be along the z-direction. The sign of the exchange integral J indicates the ferromagnetic ($J > 0$) and antiferromagnetic coupling ($J < 0$) in the system.

It is convenient to rewrite the Hamiltonian $\mathcal{H}$ by using the rising ($S_i^+ = S_i^x + iS_i^y$) and lowering spin operator ($S_i^- = S_i^x - iS_i^y$):

$$\mathcal{H} = -J \sum_{i,j} [S_i^z S_j^z + \frac{1}{2}(S_i^+ S_j^- + S_i^- S_j^+)] - H \sum_i S_i^z. \tag{2.12}$$

For $J > 0$, the ground state (the saturation magnetization) is expressed by $|0\rangle = \prod_i |S\rangle_i$, where $|S\rangle_i$ is the eigenspinor of the i-th atom and is composed of all spins oriented along the z direction (the quantization axis). Consequently, the ground state energy and the total spin in the system are expressed as

$$\mathcal{H} = E_0|0\rangle, \ E_0 = -\gamma JS^2N - HSN \tag{2.13}$$

$$S^z|0\rangle = SN|0\rangle \tag{2.14}$$

where N is the number of lattice sites, and γ is the number of nearest-neighbors on each site. The first term of E_0 is associated with the exchange interaction and the second one arises from the external magnetic field.

Considering the lowest excited state in the system, a most straightforward excited state is to flip one spin at a certain lattice site. As a result, a total angular momentum transfer of $1\hbar$ in the system is obtained, i.e., $S \rightarrow S - 1$ in Eq. (2.14). We assume that the corresponding state of a spin-flip excitation at one particular lattice site n is given by $|\downarrow_n\rangle$. Although $|\downarrow_n\rangle$ is an eigenstate of S^z, i.e., $S^z|\downarrow_n\rangle = (SN - 1)|\downarrow_n\rangle$, it is not an eigenstate of the Heisenberg Hamiltonian in Eq. (2.11) or (2.12). The reason is that once a single spin-flip at a particular site is created, the flipped spin states on its neighboring sites are possible to be excited. Therefore, instead of a single spin-flip, a lower-energy excitation of a "delocalized" state is proposed. The "delocalized" state is described by a linear combination containing of all possible single spin-flip state at different lattice sites. The normalized one-magnon eigenenergy is given by

$$|\mathbf{q}\rangle = \frac{1}{\sqrt{N}} \sum_n e^{i\mathbf{q}\cdot\mathbf{r}_n} |\downarrow_n\rangle \tag{2.15}$$

The above equation is nothing but the Bloch's theorem.

As a consequence, an eigenstate of the Heisenberg Hamiltonian and the z-component of the total spin S^z are shown below.

$$\mathcal{H}|\mathbf{q}\rangle = [E_0 + \hbar\omega(\mathbf{q})]|\mathbf{q}\rangle \tag{2.16}$$

$$S^z|\mathbf{q}\rangle = (SN - 1)|\mathbf{q}\rangle \tag{2.17}$$

where the excitation energy corresponds to $\hbar\omega(\mathbf{q})$ in Eq. (2.16). The quantum number of the z-component for the total spin S^z is $SN - 1$. Analogous to an excitation from the ground state with one spin flip, a state of $|\mathbf{q}\rangle$ can be visualized as a completely "delocalized" state in a ferromagnetic Heisenberg model. Correspondingly, a magnon is a bosonic quasiparticle with an angular momentum of $1\hbar$.

The expectation value of the local spin deviation, given by the local-site operator S_i^z, in a one-magnon state is

$$S_i^z|\mathbf{q}\rangle = (S - \frac{1}{N})|\mathbf{q}\rangle \tag{2.18}$$

The above equation clearly demonstrates that the excitation of a magnon reduces the z-component of each spin by $\frac{1}{N}$. This is the reason that the semi-classical picture of a magnon works.

By subtracting the energy of the ground state in Eq. (2.13) from Eq. (2.16), the energy of the one-magnon state ($\hbar\omega$) as a function of the wave vector $\mathbf{q}$ on the basis of the eigenstate $|\mathbf{q}\rangle$ is obtained.

$$\hbar\omega(\mathbf{q}) = H + 2JS\sum_{\mathbf{r}}[(1 - \cos(\mathbf{q}\cdot\mathbf{r})] \tag{2.19}$$

where $\mathbf{r} = \mathbf{r_i} - \mathbf{r_j}$.

For long wavelengths $|\mathbf{q} \cdot \mathbf{r}| \ll 1$ and $H = 0$, in accordance with the lattice constant $|\mathbf{r}| = a$ and $\sum_{\mathbf{r}} (\mathbf{q} \cdot \mathbf{r}) = 2q^2 a^2$,[2] the magnon dispersion relation can be simplified to

$$\hbar\omega(q) \simeq 2JSa^2 q^2 = Dq^2 \tag{2.20}$$

where D is called magnon stiffness constant. One can find that $\omega = 0$ as $q = 0$ and $H = 0$ due to the rotation invariance of the underlying Hamiltonian, in agreement with the Goldstone theorem. According to Eq. (2.20), for small q, a quadratic relation of the magnon energy to the small wave-vector q is obtained.

2.1.3   Magnons in thin ferromagnetic films

As discussed above, the theoretical descriptions of the magnon dispersion relation for a ferromagnetic chain has been derived by quasi-classical and quantum-mechanical methods. One would expect that if we extend the system from an one-dimensional chain to a two-dimensional atomic layer, for the isotropic exchange interaction in the system (i.e. $J = J_x = J_y$), the magnon dispersion relation of an one-atomic-layer ferromagnetic film is the same as that of a ferromagnetic chain. However, it is possible to break the degeneracy and exhibit multi-magnon branches when in-plane anisotropic exchange interaction is present in the system.

By piling up atomic layers along the z-axis, quantum confinement along the direction perpendicular to the film surface leads to the formation of discrete magnon states with "quantized" out-of-plane wave-vectors. For a thin ferromagnetic film, the wavelength of the confined magnon modes along the out-of-plane direction is compatible with the total phase shift of $n\pi$ among spins ($n = 0, 1, 2, ...$), which arises from a "round-trip" criterion between two boundaries of the film. They are standing spin waves.

By adding a new atomic layer to the film, an additional state with an extra half-wavelength of the standing spin wave is generated to accommodate a new atomic potential well. As a result, the number of confined magnon modes is equal to the number of atomic layers in a film. For a ferromagnetic film with N atomic layers, the standing spin waves (confined magnon modes) can be characterized explicitly by the number of n half-wavelength envelopes of standing spin waves with the number of n nodes spanned in thin magnetic films where $n = 0, 1, ..., N - 1$.

In contrast, due to the translation invariance in the film plane, magnons can propagate infinitely along any directions in the xy-plane. A set of magnon dispersion

---

[2]Here $\mathbf{q}$ is a wave-vector of a magnon and $\mathbf{r}$ is the position vector of the nearest-neighbor atoms. The factor of 2 results from the properties of the reciprocal and direct lattice. A detailed derivation can be found in Ref. [33].

relations as a function of in-plane wavevector, $\hbar\omega_n(q_\parallel)$, are obtained for the confined magnon modes with $n = 0, 1, ..., N - 1$.

The behavior of the confined magnon modes in a thin ferromagnetic film can be described by the quantum mechanical and classical Heisenberg models, in a similar manner to Sec. 2.1.1 and 2.1.2. However, to get the virtual microscopic pictures of the spin configurations among atomic layers, we introduce the theoretical description of the quasi-classical vector model as follows. For a single-element ferromagnetic film composed of N atomic layers, the Heisenberg Hamiltonian is given by

$$H = -\sum_{\alpha=1}^{N} \sum_{i,j_\parallel} J_\parallel \mathbf{S_i^\alpha} \cdot \mathbf{S_{j_\parallel}^\alpha} - \sum_{n=1}^{N-1} \sum_{i,j_\perp} 2J_\perp \mathbf{S_i^\alpha} \cdot \mathbf{S_{j_\perp}^{\alpha+1}} \qquad (2.21)$$

where α refers to a certain atomic plane of the film. The lack of a factor 2 in the first term is due to the double-counting of the exchange constant between spins within the same atomic layer. In Eq. (2.21), the exchange interaction between spins within the same atomic layer is referred to *intralayer* exchange interaction, $J_\parallel$. The one between the neighboring layers is referred to as *interlayer* exchange interaction, $J_\perp$.

Based on Eq. (2.3), for $N > 2$, the number of N torque equation is generated. The linear equations are given by

$$\hbar\omega \begin{pmatrix} A_1 \\ . \\ . \\ . \\ A_N \end{pmatrix} = \mathbf{I_N} \begin{pmatrix} A_1 \\ . \\ . \\ . \\ A_N \end{pmatrix} \qquad (2.22)$$

where $\mathbf{I_N}$ is a $(N \times N)$ coefficient matrix including the intra- $(J_\parallel^N)$ and inter-layer $(J_\perp^N)$ exchange constants of each atomic layer. The eigenvector consists of the magnon amplitude of each atomic layer, referred to as $A_1, A_2, ..., A_N$. The magnon energy $\hbar\omega$ is the eigenvalue of the coefficient matrix $\mathbf{I_N}$. As a result, the number of N magnon dispersion relations, $\hbar\omega_n(q_\parallel)$, is constituted. By comparing the measured and calculated magnon dispersion relation of all magnon modes, the layer-dependent intra- $(J_\parallel^N)$ and inter-layer $(J_\perp^N)$ interatomic exchange parameters in a thin ferromagnetic film can quantitatively be determined.

An example of a three-atomic-layer ferromagnetic film

The aim of this work is to investigate the magnon dispersion relation of 3 ML Co films grown on different substrates. Therefore, in this section the focus is put on the confined magnon modes in a three-atomic-layer ferromagnetic film ($N = 3$). Figure 2.2 shows a "snapshot" of the standing spin waves of such a free standing film. The

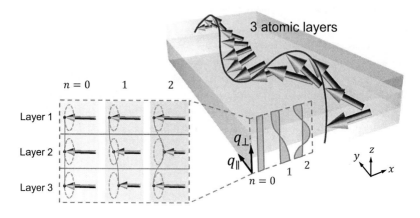

Figure 2.2: The schematic diagram of the confined magnon modes with the quantization number $n = 0$, 1 and 2 for a 3 ML ferromagnetic free-standing film. The modes $n = 0$, 1 and 2 correspond to zero, one and two nodes of standing spin waves inside the film, respectively. The number of zero, one and two half-wavelength envelopes of standing spin waves inside the film is shown. The "quantized wave-vector" perpendicular to the film surface satisfies the condition of $q_\perp = \frac{n\pi}{d}$, where d is the thickness of films and n is the quantum number.

layer-resolved oscillating spins of three magnon modes corresponding to the quantum number of n are depicted. The $n = 0$ (zero nodes) mode indicates the coherent spin precession. The $n = 1$ mode corresponds to one node in the center of the film, and the $n = 2$ mode corresponds to two nodes inside the film. Correspondingly, for magnon modes of $n = 0, 1$, and 2, the number of zero, one, and two half-wavelength envelopes of standing spin waves inside the film are demonstrated, respectively.

The presence of the standing spin waves reflects that the spins between adjacent layers are no longer aligned parallel to each other. Instead, a rather large deviation of the precession phase of the spins is found. The highest-energy mode is characterized by the largest deviation of the precession phase between adjacent layers. Thus, the energies of standing spin waves increase as the quantum number n increases.

The wave-vector of magnons parallel and perpendicular to the film surface is referred to as $q_\parallel$ and $q_\perp$. Due to the the broken translational symmetry normal to the surface, the allowed values of $q_\perp$ are given by the film thickness, such that they satisfy the condition $q_\perp = \frac{n\pi}{d}$, where d is the thickness of films and n is the integer-valued quantum number.

As shown in Fig. 2.3, the continuous dispersion relation of the bulk at $q_\parallel = 0$ (the orange line) is substituted by three discrete states of the confined magnon modes with $n = 0, 1$ and 2 for a 3 ML film. The "quantized wave-vector" perpendicular to the

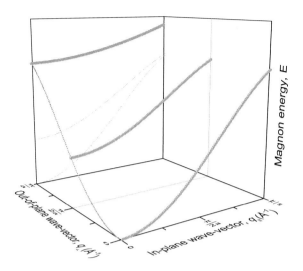

Figure 2.3: The schematic magnon dispersion relation of magnon modes of $n = 0, 1$ and 2 for a 3 ML ferromagnetic film. The continuous dispersion relation of the bulk (the orange line) at $q_\parallel = 0$ is substituted by three discrete states of the confined magnon modes with $n = 0, 1$ and 2 at $q_\perp = 0$, $\frac{\pi}{2a}$ and $\frac{\pi}{a}$ where a is the interlayer distance of the film. Away from the surface normal to the surface plane, owing to the transitional invariance in the film plane, each magnon mode has its dispersion relation associated with the in-plane wave-vector $q_\parallel$ over the whole surface Brillouin zone (the blue lines).

film surface in a 3 ML film is $q_\perp = 0$, $\frac{\pi}{2a}$ and $\frac{\pi}{a}$, where a is the interlayer distance of the layers. However, away from the surface normal to the surface plane, owing to the transitional invariance in the film plane, each magnon mode has its dispersion relation associated with the in-plane wave-vector $q_\parallel$ centered about the $\overline{\Gamma}$-point over the whole surface Brillouin zone (the blue lines in Fig. 2.3).

In SPEELS measurements, one scans the energy-loss while keeping the in-plane wave-vector transfer fixed during the scan (so-called constant-q scan), indicated by the vertical arrows in Fig. 2.4a. It is expected that for a 3 ML film, three magnon peaks are observes in the energy-loss spectra, denoted by three green circles in Fig. 2.4a, as an example of the SPEELS difference spectra recorded on 3 ML Co/Ir(001) at an in-plane wave-vector transfer of $\triangle K_\parallel = 0.5$ Å^{-1} along the $\overline{\Gamma} - \overline{M}$ direction (Fig. 2.4b). Three magnon peaks in energy-loss spectra are observed, which are fitted well by the three Lorenzian lineshapes. The blue, green and pink curves in Figs. 2.4a and b correspond to the confined magnon modes of $n = 0$, 1 and 2, respectively. Remarkably, with precise film thickness in terms of the number of atomic layers, the characteristics

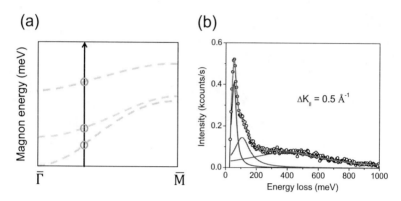

Figure 2.4: (a) The schematic magnon dispersion relation for a 3 ML ferromagnetic film. In SPEELS measurements, the energy-loss is scanned while the in-plane wave-vector transfer is kept constant (constant-q scan) (the vertical arrow). Three magnon peaks in energy-loss spectra are expected (three green circles). (b) An example of the SPEELS difference spectra recorded on 3 ML Co/Ir(001) at an in-plane wave-vector transfer of $\triangle K_\| = 0.5$ Å^{-1} along the $\overline{\Gamma} - \overline{M}$ direction. Three magnon peaks are fitted by three Lorenzian lineshapes. Blue, green and pink curves in (a and b) indicate the confined magnon modes of $n = 0$, 1 and 2 for a 3 ML film, respectively.

of the confined magnon modes and the corresponding magnon dispersion relation in ultrathin magnetic films can be studied by SPEELS. The effect of the relative strength between the intra- and inter-layer exchange interaction on the dispersive feature of the confined magnon modes will be discussed in detail in Sec. 5.2.

2.2 Stoner excitations in itinerant-electron ferromagnets

In addition to the collective spin-wave excitations, another magnetic excitations in an itinerant ferromagnet, resulting in a reduction of the magnetization by $1\hbar$, are single-particle Stoner excitations. In Fig. 2.5a, a schematic representation of the Stoner excitation with a transfer of a total angular momentum of $\hbar$, wave-vector $\mathbf{q}$, and energy ε is shown. The density of Stoner states is determined by the available electronic transitions between the occupied majority spin states ($\uparrow$) with crystal wave-vector $\mathbf{k}$ and the empty minority spin states ($\downarrow$) with wave-vector $\mathbf{k} + \mathbf{q}$. The continuum of the Stoner excitations is shown as the shaded area in Fig. 2.5b.

In particular, at $q = 0$, electron and hole of opposite spin have the same wave-vector $\mathbf{k}$ in the Brillouin zone. The energy of this electron-hole pair is given by the

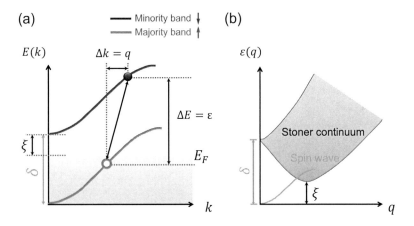

Figure 2.5: (a) Schematic representation of the Stoner excitations with an electron-hole pair of opposite spin. An electron is excited from an occupied majority-spin state to an empty minority-spin state. (b) The Stoner continuum (the shaded area) and collective spin-wave excitations (the green line) coexist in an itinerant ferromagnet. At $q = 0$, the Stoner excitations carry the energy δ, which corresponds to the "average" exchange splitting of majority- and minority-spin states. The minimum energy to excite a Stoner electron-hole pair is ξ, corresponding to the energy between the Fermi level and the bottom of the minority spin band.

"average" exchange splitting δ of majority- and minority-spin states. As a result, a Stoner excitation requires an energy of δ at $q = 0$ (Fig. 2.5b). Moreover, the minimum energy to excite a Stoner electron-hole pair is ξ, the so-called "Stoner gap." This energy corresponds to a distance from the Fermi level to the bottom of the minority spin band. Note that the Stoner continuum depends strongly on details of the electronic structure. We only discussed the simple case of the homogeneous three-dimensional electron gas with "parabolic" bands.

The dispersion relation of the "acoustic" magnon mode ($n = 0$) is also presented in Fig. 2.5b. Particularly at higher energies, magnons decay into the Stoner continuum. The hybridization with the single-particle Stoner excitations have the consequence of broadening magnon peaks and a reduction of magnon lifetime.

For an ultrathin magnetic film, the electronic states of the substrate may contribute as extra decay channels for magnons. This effect results in a strong damping of the high-energy confined magnon modes. As a rule of thumb, the magnon lifetime decreases as their energy increases. Therefore, the damping of magnons depends strongly on the details of the interfacial electronic hybridization between films and substrates, as will be discussed in Sec. 5.5.

2.3   Electron scattering

2.3.1   Low-energy inelastic electron scattering

For the low primary energy electrons (in the range between several eV and a few keV), scattered electrons are measured in the reflection geometry from a solid surface, due to their rather short penetration length. For instance, the mean free path of electrons with 10 eV above the vacuum level in Co is approximately 6.5 (8.5) Å for minority- (majority-) spin electrons [34, 35]. In other words, the low-energy electrons can penetrate and interact at most three or four interatomic spacing in solids. Due to the fact that low-energy electron spectroscopy is highly surface sensitive, measurements under ultrahigh vacuum conditions are required.

For reflection-mode high-resolution spectroscopy, a monochromator, by using a hemispherical or a cylindrical electron analyzer, can reduce the width of the energy distribution of primary beams to a few meV. Similarly, the energy of scattered electrons at a certain angle is measured using an analyzer with electron optics similar to the monochromator, but in the reverse order. Combining these two sub-systems, an energy resolution down to around 1 meV can be achieved. As a result, low-energy inelastic electron scattering techniques sufficiently resolve the low energy-loss signals about vibrational modes from adsorbed atoms (or molecules) and elementary excitations (phonons, plasmons, magnons, electron-hole pair excitations, *etc.*) in ultrathin films.

In particular, inelastic scattering processes involving the excitation (the annihilation) of phonons or magnons bring about very small energy losses (gains) (< 100 meV). The energy-loss signal contributed from these excitations cannot be distinguished from a zero energy-loss peak without using high-resolution electron analyzers. Correspondingly, this technique is usually referred to high-resolution electron energy-loss spectroscopy (HREELS).

In general, low-energy electrons interact with a sample surface involving three different types of inelastic scattering mechanisms: dipole, impact and resonance scattering [36]. Each type of energy-loss mechanism is characterized by its own particular scattering features. They are categorized based on several factors, such as interaction between incident electrons and crystal surfaces, angular distributions of scattered electrons, scattering cross sections, and applications of selection rules. To capture the information regarding low-energy excitations in low-dimensional systems, electron scattering experiments with a short mean free path of electrons in solids are the most versatile techniques in surface science, in contrast to photons and neutrons as a probe.

Long-range dipole scattering

The origin of the dipole scattering is due to the moving incoming electrons that interact via their accompanying electric field with the change-density fluctuation of elementary excitations or the oscillating electric dipole moment of a molecule at the surface. Due to the relatively long-ranged dipole-dipole interaction (e.g., of the order of 100 Å), incident electrons are able to excite vibrational excitations of molecules.

The angular distribution of scattered electrons via electric dipole scattering is very narrow and is mainly confined to the so-called dipole lobe around the specular reflected beam [37]. This results in only very small wave vector transfers ($\leq 10^{-2}$ Å^{-1}) in the dipole scattering regime. The description of excitations originating from dipole scattering mainly relies on the dielectric function [37], which is related to the macroscopic properties of the target. Apart from the long-range dipole scattering, there is another mechanism at large scattering angles, i.e., impact scattering.

Short-range impact scattering

In contrast to the dipole scattering, the impact scattering arises from the short-range Coulomb interaction (of the order of 1 Å) between incident electrons and the sample surface. Owing to the conservation of parallel-momentum in the sample surface, the large-angle impact scattering is responsible for large momentum transfers. Thus, high wave-vector (short wavelength) excitations can be explored by impact inelastic scattering. In the impact scattering process, the incident electrons reach very close to the surface and even penetrate into the solid. This also gives rise to electron exchange, since electrons are indistinguishable and their wavefunctions overlap. Therefore, the impact scattering is accompanied by electron-exchange. This is the most important mechanism for probing magnetic excitations in ultrathin films.

In a simplified kinematic picture, the inelastic cross section in the dipole dominated region can be described by a macroscopic theory (the phenomenological approach), while in the impact scattering region one requires an adequate description based on the microscopic approaches. The cross section in the impact scattering is substantially smaller (about 2−3 orders) than that of the dipole one [38]. It is important to note that, in the measured electron energy-loss spectra, contributions of both the dipole and impact scattering coexist.

2.3.2   Spin-dependent effects in elastic electron scattering

In general, there are two main relevant spin-dependent interaction in the scattering process that cause the spin polarization of elastically scattered electrons from a crystal surface. They are exchange interaction and spin-orbit interaction, which are essential mechanism exploited in electron spin detectors. Both interactions can be simultaneously present in the system.

The physical concept of spin dependence in elastic scattering can be most simply visualized for the case of the electron scattering from a free atom. One can consider the unpolarized incident beam as a composition of two equal fractions with opposite spin directions, being perpendicular to the scattering plane. The scattering potential consists of the electrostatic and spin-orbit interaction potential, i.e., $U = U_0 + U_{SO}$. The spin-up and spin-down electrons with the same trajectories experience a different total scattering potential as a consequence of U_{SO}. In a nonrelativistic limit, the spin-orbit interaction potential can be described by

$$U_{SO} = \frac{1}{2m^2c^2}\frac{1}{r}\frac{dV}{dr}(\mathbf{L}\cdot\mathbf{S})$$
(2.23)

where $\mathbf{L}$ is the angular momentum with respect to the scattering center and its vector is defined by the wave vectors of the incident and the scattered electrons [39]. $\mathbf{S}$ is the spin of the incident electron. Since U_{SO} contains the scalar product of $\mathbf{L}\cdot\mathbf{S}$, for elastic electron-nucleus scattering, the coupling of same angular momentum of electrons with the opposite spin orientation leads to additional scattering potential being added to or subtracted from the (screened) Coulomb potential. For a Coulomb potential of $V(r) = -\frac{Ze^2}{r}$, $U_{SO} \propto \frac{Z}{r^3}(\mathbf{L}\cdot\mathbf{S})$ is obtained. As a result, the effect of spin-orbit interaction is the strongest for the non-magnetic heavy elements such as W, Pt or Au due to $U_{SO} \sim Z$ and it is largest in the vicinity of the nucleus due to $U_{SO} \sim \frac{1}{r^3}$ [40]. The different sign of U_{SO} results in spin-up and spin-down electrons being scattered preferentially to the right- and left-hand wise of the nucleus, respectively [39]. This leads to a "left-right asymmetry" in the polarization of the scattered electrons. Thus, a scattering asymmetry through the spin-orbit interaction (A_{so}) is found. The spin-orbit asymmetry is only sensitive to the spin polarization components perpendicular to the scattering plane.

Using a ferromagnetic target,[3] the differential scattering cross section of the elastic exchange scattering depends on the relative orientation of the polarization vectors for the primary electron beam ($\mathbf{P}$) and the sample magnetization ($\mathbf{M}$). For the energy range we studied, the spin-spin dipole interaction between electron beam and target can be neglected. As a result, the measured normalized asymmetry (A_{ex}) between the two intensities of the opposite beam polarization is proportional to the $\mathbf{P}\cdot\mathbf{M}$ term [39]. This indicates the exchange scattering asymmetry is sensitive to the polarization component along the magnetization axis of the target. As a consequence, one can obtain asymmetries for spin-polarized scattered electrons caused by spin-orbit

---

[3]Note that for non-magnets, during the scattering process the exchange interaction is also involved, but their contribution is the same for spin-up and spin-down electrons [40].

or exchange interaction.

Chapter 3

Experimental techniques

3.1 Spin-polarized electron energy loss spectroscopy

3.1.1 Spin polarized inelastic electron scattering

In this section we discuss the spin-dependent energy losses of a primary electron during an inelastic scattering process. We recall that the impact scattering process arises from the short-range Coulomb interaction between the incident electrons and the sample surface. In contrast to the long-range dipole scattering mechanism, incident electrons penetrate into the solids in impact scattering regime and thus energy loss events can occur within few atomic layers at the solid surface. Two fundamental magnetic excitations, collective spin waves and Stoner excitations, in ferromagnetic surfaces can be investigated by means of SPEELS. In the following, we consider only the electron-hole excitations to interpret possible spin-dependent inelastic scattering events in ferromagnetic materials.

Figure 3.1 shows possible processes associated with the generation of an electron-hole pair. The processes take place during the inelastic scattering of an incoming electron with a kinetic energy E_i, scattered back from the ferromagnetic surface with a kinetic energy of $E_i - \varepsilon$. Here, only the incoming minority-spin (spin-down) electrons are presented. In all scattering processes in Fig. 3.1 the energy loss ε of the incident and scattered electrons is the same as the energy (ε) necessary to create an electron-hole pair separated by the Fermi level. The spin is conserved during the transition. Therefore, only transitions from the states with the same spin orientation are allowed. The energy loss processes are classified into the "non-flip" and the "flip" scattering processes.

In Fig. 3.1a-c, an incoming spin-down electron experiences three energy-loss processes in which the incident and scattered electrons have the same spin orientation. All three processes are associated with the generation of an electron-hole pair of the same spin across the Fermi level. They are referred to as the "non-flip" process. This

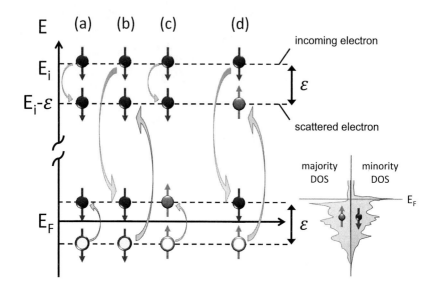

Figure 3.1: Schematic interpretations of inelastic electron scattering processes associated with electron-hole pair excitations for incident spin-down electrons (the minority-spin electrons in ferromagnets). All processes take place in the inelastic scattering of an incoming electron with a kinetic energy E_i and scattered from the ferromagnetic surface with a kinetic energy of $E_i - \varepsilon$. (a-c) The processes in which the incident and scattered electrons have the same spin character are denoted as "non-flip" processes. (d) The spin of the scattered electron is opposite to the one of the incident electron. This process is referred to as "flip" scattering process. The scattering process of (d) results in the creation of an electron-hole pair of the opposite spin, so-called Stoner excitations.

process is the same as optical absorption experiments according to the dipole selection rule. Another case is the "flip" scattering process (Fig. 3.1d) in which the spin of the scattered electron is opposite to the one of the incident electron. In this case, an electron-hole pair of the opposite spin is created. These are the Stoner excitations carrying spin of $1\hbar$.

Note that, in the "flip" scattering process, a spin reversal seems to take place when only looking at the initial and final states of the electron beams. However, in reality none of the spins is flipped. Similarly, the "non-flip" and "flip" scattering processes during the electron energy loss events also take place for an incoming majority-spin (spin-up) electron. An extended discussion can be found in Ref. [40]. In the scattering of electrons from both magnetic and non-magnetic surfaces, the electron exchange events always exist. However, in non-magnetic surfaces, the scattered electrons remain a net balance of spins, due to the equivalent amount of spin-up and spin-down electrons in the system.

The relative scattering rate regarding these processes and their difference in spectral intensity have been experimentally measured and discussed in Refs. [40–43]. In contrast to the similar intensity among the "non-flip" scattering channels, it has been found that in the "flip" processes the energy-loss intensity for the incident minority electrons is higher than that of the incident majority electrons [41,42]. In other words, Stoner excitations with a majority-hole and minority-electron are much more likely to take place than a minority-hole and majority-electron. The exchange splitting of the $3d$ bands in ferromagnets leads to the fact that the empty density of states for the minority-spin are larger than the ones for majority-spin. Therefore, the incident minority-spin electrons have a larger probability for the "flip" excitations. This results in a quite large asymmetry in the measured SPEELS spectra in which the magnetic excitations are excited.

3.1.2   Principles of SPEELS

The energy loss of electrons during the scattering process is equivalent to the energy of the excitation left behind in the sample. In addition, due to the conservation of total angular momentum during the scattering process, the change of the angular momentum of incoming and scattered electrons leads to a change with the same magnitude but opposite direction of the angular momentum in the sample. An excitation of a magnon is analogous to one spin flip from majority to minority spin. Therefore, creation of a magnon carrying total angular momentum of $1\hbar$ in ferromagnets is only possible when the incoming electron is of minority-spin ($S = -\frac{1}{2}\hbar$) and the scattered electron is of majority-spin ($S = +\frac{1}{2}\hbar$) via an electron exchange scattering process. In such a scattering process, the total angular momentum in the sample is reduced by $\hbar$ which corresponds to exciting a magnon. Due to the spin-dependent nature of magnons, this mechanism leads to the essential consequence that magnon peaks are distinguished from other kinds of excitations in SPEEL spectra.

Vice versa, the annihilation of a magnon in ferromagnets is only possible when the incoming electron is of majority-spin and the scattered electron is of minority-spin. As a result, a magnon peak associated with the magnon annihilation process can be seen in the energy gain region. The ratio between energy gain (I_{gain}) and loss (I_{loss}) intensity follows the Boltzmann factor, i.e. $\frac{I_{gain}}{I_{loss}} = e^{(-\hbar\omega/kT)}$, depending on the temperature of the sample. In this study, we demonstrate only the excitation of magnons in the loss region of the SPEEL spectra since we find the same energy for the created or annihilated magnons in the respective energy loss and gain region.

Note that the scattering cross section of magnetic excitations strongly depends on the energy of the incident electrons. For instance, the higher intensities of the single-

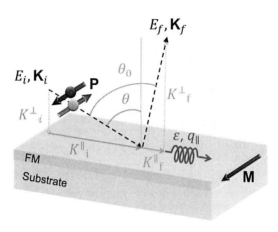

Figure 3.2: The scattering geometry in SPEELS experiments. θ_0 is the angle between the directions of incident and scattered electron beams. θ is the angle between the incident beam direction and the surface normal. The energy and wave-vector of the incident and scattered electrons are denoted as E_i, $\mathbf{K_i}$ and E_f, $\mathbf{K_f}$, respectively. The energy and the in-plane wave-vector of magnetic excitations are denoted as ε and $q_{\parallel}$, respectively.

particle Stoner excitations is observed as the incident energy above 25 eV [44,45]. On the other hand, the pronounced magnon peak in the loss spectrum is observed only at a primary energy below 10 eV [29].

According to the conservation of energy and in-plane momentum, the energy difference between the incident (E_i) and scattered electrons ($E_f = E_i - \varepsilon$) must be equal to the energy of the excitations ($\varepsilon = \hbar\omega$) left behind in the sample. The energy ε and the in-plane wave-vector $q_{\parallel}$ of the excitation in a given scattering geometry is described by

$$\varepsilon = \hbar\omega = E_i - E_f \tag{3.1}$$

$$q_{\parallel} = -\Delta K_{\parallel} = K_{\parallel}^i - K_{\parallel}^f = \mathbf{K_i}\sin\theta - \mathbf{K_f}\sin(\theta_0 - \theta) \tag{3.2}$$

where $\mathbf{K_i}$ and $\mathbf{K_f}$ indicate the wave-vector of the incident ($\mathbf{K_i}$) and the scattered ($\mathbf{K_f}$) electron beams in the SPEELS experiments, respectively (see Fig. 3.2). θ_0 is the angle between the directions of the incident- and the scattered-electron beam, also being the angle between the monochromator and the analyzer. θ is the angle between the incident electron beam and the surface normal. θ_0 is fixed at 80° in this study.

By changing the angle between the incident and scattered beam, the desired wavevector can be achieved with an in-plane momentum resolution of 0.03 ± 0.01 Å^{-1}. Thus, one can obtain the magnon energy (ε) as a function of the in-plane wave-vector ($q_{\parallel}$). The measurements for probing magnons at a given positive and negative

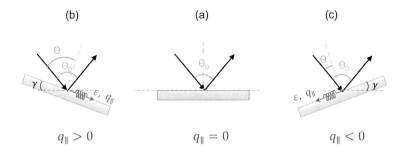

Figure 3.3: A schematic representation of the scattering geometry in SPEELS experiments at zero (a), positive (b) and negative (c) wave vectors. θ_0 is the angle between the directions of incident and scattered electrons. θ is the angle between the incident beam and the surface normal. γ is the rotation angle of the sample. (a) $\gamma=0$, specular scattering geometry. (b) $\gamma<0$, off-specular scattering geometry for probing magnons with positive in-plane wave-vector. (c) $\gamma>0$, off-specular scattering geometry for probing magnons with negative in-plane wave-vector.

wave-vector are feasible, as shown in Fig. 3.3. In SPEELS measurements, magnons are probed in the system at a positive (Fig. 3.3b) and a negative (Fig. 3.3c) wave vectors, corresponding to the rotation of the angle γ in a clockwise and a counterclockwise direction with respect to the specular geometry (Fig. 3.3a).

3.2 The experimental setup

3.2.1 The ultra high vacuum system

All experiments in this work are performed in an ultrahigh vacuum (UHV) system with a base pressure in the low 10^{-11} mbar range. As shown in Fig. 3.4, our experimental set-up is composed of three main chambers. They are the sample preparation, GaAs-photocathode preparation, and SPEELS chambers, respectively.

The sample preparation chamber is equipped by a number of standard in situ sample preparation and characterization instruments. The samples are grown by e-beam evaporators for molecular beam epitaxy (MBE). Their chemical, structural, and magnetic properties are examined by low (medium) energy electron diffraction [L(M)EED], Auger electron spectroscopy (AES), and longitudinal magneto-optic Kerr effect (LMOKE).

The cathode chamber is used to prepare the GaAs-photocathode which is used as a source of spin-polarized electrons in the SPEELS measurements. It is equipped with two Cs-dispensers, a leak valve for O_2 gas, a heating facility, and a laser with 670 nm-wavelength and 10 mW power.

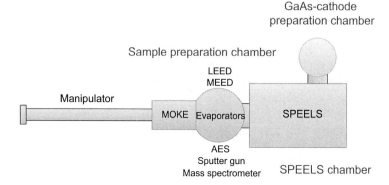

Figure 3.4: The top view of experimental set-up in our UHV system. Three main chambers are represented by different colors. They are the sample preparation (orange), GaAs-cathode preparation (green), and SPEELS chambers (blue). For the sample analysis chamber, standard sample characterization instruments are denoted. Samples can be transferred between the analysis chamber and the SPEELS chamber by the manipulator. In the GaAs-cathode preparation chamber, the spin-polarized photocathode is prepared by sequentially evaporating Cs and O on a clean GaAs surface. The GaAs-photocathode can be transferred between cathode preparation chamber and SPEELS chamber by a UHV wobble stick. The spin-resolved electron scattering experiments are performed in the SPEELS chamber.

The SPEEL-spectrometer is integrated in the SPEELS chamber. Before performing the SPEELS experiments, two things are needed to be prepared. The film on substrates is characterized and magnetized in the preparation chamber. Meanwhile, the proper photocurrent from the strained GaAs photocathode is obtained. As both materials are prepared, they are transferred to SPEELS chamber.

3.2.2   SPEEL-spectrometer

The layout of the SPEEL-spectrometer equipped with the spin-polarized GaAs-photocathode is shown in Fig. 3.5. The reversal of the spin-polarization of the photoexcited electrons from the GaAs photocathode is achieved by switching between right ($\sigma+$) and left ($\sigma-$) circularly polarized light. In the SPEELS experiments, the spectrometer is used to analyze the intensity of scattered electrons from a sample surface under a given scattering geometry. By measuring intensity changes upon reversal of the beam polarization, information about magnons excited in the system can be obtained effectively, being able to distinguish magnetic excitations from phonon and vibrational excitations on the sample surface.

Since only incident electrons with minority spins can excite magnons, the maxi-

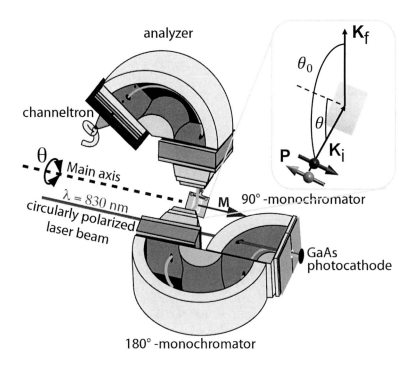

Figure 3.5: The layout of the SPEEL-spectrometer equipped with the spin-polarized GaAs-photocathode. The emitted electrons from GaAs surfaces pass a 90° pre-monochromator and an 180° monochromator with electrostatic deflection. The scattering plane is chosen to be perpendicular to the spin orientation of the incident electrons. The total scattering angle is fixed at 80° in this study. The electrons scattered from the surface are then analyzed by an energy analyzer, which is a standard EELS-monochromator with a deflection angle of 146°. The simplified electron trajectory is shown as the blue line. The schematic diagram of the spectrometer is adapted from Ref. [46]. The inset denotes the wave-vector of the incident ($\mathbf{K_i}$) and the scattered ($\mathbf{K_f}$) electron beams in SPEELS experiments. The monochromatized incident electron beam has the spin polarization parallel and antiparallel to the main axis of sample rotation (around the angle θ), being perpendicular to the scattering plane.

mum asymmetry (or maximum difference intensity) in a SPEELS spectrum is obtained for two incident electron beams with opposite spin orientation along the magnetization axis of the sample. For this purpose, a 90° deflector as a pre-monochromator is combined with a 180° deflector as a main monochromator. The spin-orientation of emitted electrons from the GaAs-photocathode is along the emission direction.

After passing through the monochromator, the monochromatized incident electron beam has its spin polarization parallel and antiparallel to the main axis of sample rotation (around the angle θ), being perpendicular to the scattering plane (see the inset in Fig. 3.5). The energies of the scattered electrons are analyzed by a 146° deflector. The energy resolution is 10-20 meV. More details of the SPEEL-spectrometer can be found in Ref. [31].

In general, the intensity (the number of scattered electrons) as a function of the loss (or gain) of energy is represented in the SPEEL spectrum. At each energy-loss (or -gain) position, the intensities of scattered electrons for incident electrons of opposite polarization are recorded. I_- and I_+ spectra refer to the measured intensities when the spin polarization of the incoming electron beam is parallel and antiparallel to the main axis of sample rotation, respectively.

3.2.3   The GaAs-photocathode

In SPEELS measurements, a GaAs-photocathode is used to obtain spin-polarized electrons by photoemission with circularly polarized light. However, electrons excited are so far unable to escape from the crystal surface with its high electron affinity. This is due to the fact that the energy position of the vacuum level lies around 4 eV above the bottom of the conduction band. Fortunately, by treatment with Cs and O_2 on the crystal surface, a stable negative electron affinity GaAs-photocathode can be obtained. In such case, electrons at the bottom of the conduction band can escape into the vacuum without paying additional energy and destroying their spin orientation. This treatment is usually called the "yo-yo" procedure as shown in Fig. 3.6.

All the procedures of cleaning, activating, and maintaining the GaAs photocathode require ultrahigh vacuum. The procedures to prepare the GaAs-photocathode are described as follows.

(i) To obtain a clean and smooth GaAs surface, the GaAs is slowly heated to about 620 K within 1 hour. Then, the heating power is increased to about 25 W for 5 minutes and the temperature reaches approximately 850 K. After heat treatment, the GaAs is left to cool down to about 370 K.

(ii) For substantially reducing the electron affinity of GaAs, the GaAs is exposed in sequence to Cs and O_2. Simultaneously, the photocurrent is measured by illuminating

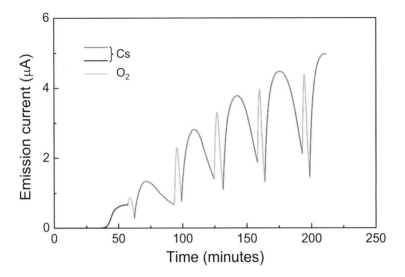

Figure 3.6: Measurement of the photocurrent of a GaAs-photocathode during the preparation by illuminating the cathode with a laser of 670 nm-wavelength and 10 mW power. The Cs exposure is continuously carried out during the whole preparation. Blue peaks indicate the O_2 exposure in the pressure of around 1×10^{-8} mbar.

the GaAs with a laser of 670 nm-wavelength and 10 mW power, as shown in Fig. 3.6. In the beginning of the Cs exposure, the photocurrent is gradually increased until it is saturated at around 0.7 μA (the red curve). Then, the adsorbed Cs on the cathode surface is subsequently oxidized by O_2 exposure at a pressure of 1×10^{-8} mbar. This leads to a sharp increase of the photocurrent. When the current drops to $\frac{1}{3}$ of the maximum value of the blue peaks, the O_2 exposure is stopped. The photocurrent increases to a higher maximum value and starts decreasing again by the continuous Cs supply. At the half of the maximum value of the orange peaks, the O_2 exposure is started again. To optimize the cathode performance, the above procedure is repeated around 5 to 8 times. At the end, the Cs exposure is stopped as soon as the maximum of the photocurrent is reached.

In the first deposition of Cs, the adsorbed Cs on the clean and annealed GaAs surface already reduces the electron affinity (Fig. 3.6). The subsequently oxidized Cs layers result in an effectively lowered GaAs vacuum level below the conduction band minimum, i.e., negative electron affinity. After the preparation procedure, the photocurrent reaches a value of about 5 μA when the GaAs-photocathode is illuminated with a laser of 670 nm-wavelength and 10 mW power.

A laser of 670 nm-wavelength is used here to monitor the photocurrent during the exposure procedures of Cs and O_2. However, to achieve both high polarization and high yield of the spin-polarized electron beam from the GaAs-photocathode, an infrared light source with the photon energy of 1.49 eV is used in SPEELS measurements. The maximum photocurrent up to 30 μA can be obtained when illuminating the GaAs cathode with a wavelength of 830 nm (photon energy of 1.49 eV) and $\approx$ 100 mW laser beam in front of the entrance slit of 90°-monochromator.

As outlined in Sec. 2.3.2, the degree of polarization of electrons emitted from the GaAs photocathode is calibrated by elastic electron scattering by means of the SPEEL-spectrometer. Using spin-dependent low energy electron diffraction from a clean W(110) surface as a spin analyzer, we obtain a spin polarization of the emitted photo electrons of about 79%.

Due to the gradual increase of the amount of contamination on the cathode surface over time, the photocurrent reduces. To obtain a stable emission of the photocurrent, we typically set the photocurrent to 1μA during the SPEELS measurements by decreasing the laser power. To compensate the decay of the photocurrent, the laser power is gradually increased over time. After the current drops to a value lower than 0.5μA at maximum laser power, the procedures of (i) and (ii) for cleaning and activating the GaAs photocathode have to be carried out again. The GaAs photocathode can be used on average for five days in a vacuum in the low 10^{-11} mbar range, before the preparation has to be repeated.

Note that a "normal" GaAs cathode provides only a theoretically limited polarization of 50% (only around 40% is actually obtained) [47]. By using superlattices or strained GaAs layers on a suitable substrate, a significant increase in the photoelectron spin polarization to more than 85% can be achieved [48,49]. The principle and a more detailed comparison have been experimentally demonstrated and discussed in Refs. [47–51].

Chapter 4

Results

4.1 The Co/Ir(001) system

4.1.1 Sample preparation and characterization

To grow the Co films on Ir(001) surfaces, the primary important step is to prepare a clean Ir surface. For cleaning the Ir(001) surface, a cleaning procedure has been developed in our laboratory [52]. The cleaning procedure consists of a few cycles of low-power flashes in oxygen atmosphere ($P_{O_2} = 6 \times 10^{-8}$ mbar) and a subsequent high-power flash. Through performing this procedure, surface contamination, especially carbon and oxygen, can be effectively removed from the Ir surface. The cleanliness of the Ir surface is checked by LEED, AES and EELS.

Two Ir(001) substrates are used in this study to investigate the magnon dispersion relation along different high symmetry directions ($\overline{\Gamma} - \overline{X}$ and $\overline{\Gamma} - \overline{M}$ of the surface Brillouin zone). One has its length and width along the Ir[110] and Ir[1$\overline{1}$0] direction and the other are along the Ir[100] and Ir[0$\overline{1}$0] direction, respectively (Fig. 4.1a).

Instead of the (1 × 1) symmetry of a bulk fcc Ir crystal, a clean Ir(001)-(5 × 1) surface is observed by LEED (Fig. 4.1). The real-space ball model of the Ir(001)-(5 × 1)-hex surface proposed by STM and IV-LEED studies is shown in Fig. 4.1b [53]. This model is characterized by the hexagonal-close-packed arrangement at top layer atoms, being 20% denser than the bulk layers beneath [53]. It is expected that the appearance of either (1 × 5) or (5 × 1) reconstructions has an equivalent possibility based on the symmetry of the cubic lattice. However, different proportions of (1×5) and (5 × 1) reconstructions may arise due to broken symmetry, especially at the edges of the Ir crystals.

Two reconstructed configurations, (5×1)-hex and (5×1+1×5)-hex are simultaneously observed on both Ir(001) crystals, as shown in Fig. 4.1c-f. The (5×1+1×5)-hex surface reconstruction is observed as two mutually orthogonal domains. In both crystals, these domains are mainly located at the central regions of the crystal. In the

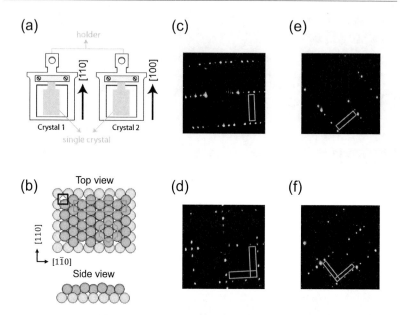

Figure 4.1: (a) A schematic representation of two Ir(001) crystals utilized in this work. The crystal 1 has its length and width along Ir[110] and Ir[1$\bar{1}$0] directions and the crystal 2, along Ir[100] and Ir[0$\bar{1}$0] directions, respectively. (b) Top and side views of the ball-model representation of a clean Ir(100)-(5 × 1)-hex surface. The (5 × 1) surface unit-cell (orange, dark-blue balls) with respect to the (1 × 1) bulk unit-cell (black, light-blue balls) and the surface hexagon (green) are indicated, which were proposed by STM and IV-LEED studies [53]. LEED patterns are taken from (c, d) crystal 1 and (e, f) crystal 2 with (c, e) (5 × 1) and (d, f) (5 × 1 + 1 × 5) superstructure spots, respectively. The corresponding (5 × 1) surface unit-cell of diffraction spots in the reciprocal space is denoted by the orange rectangle. All the LEED patterns are recorded at the primary beam energy 100 eV.

regions close to the edges a single domain structure is observed, consisting of only (1 × 5) or (5 × 1) symmetry.

Co was deposited at room temperature *in situ* by molecular beam epitaxy from a high-purity Co rod heated by electron bombardment [54]. The deposition rate and the film thickness are calibrated by means of MEED. The intensity of the diffracted beam with an energy of 3 keV is monitored during the deposition of Co on a clean Ir(001)-(5 × 1)-hex surface at room temperature. A clear intensity oscillation during the deposition is observed as shown in Fig. 4.2. This oscillation can be explained by the periodically varied surface coverage of the growing Co adatoms during the deposition. An oscillation of intensity in a periodic manner reflects the layer-by-layer growth of Co on Ir(001). Each full period in MEED spectra corresponds to the deposition one Co

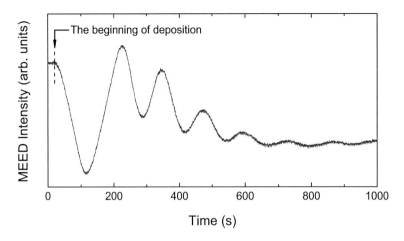

Figure 4.2: Intensity of the diffracted 3 keV grazing incidence beam as function of the deposition time of Co on a clean Ir(001)-(5 × 1)-hex surface at room temperature. An oscillation of intensity in a periodic manner reflects the layer-by-layer growth of Co on Ir(001). A complete period corresponds to the formation of one atomic Co layer.

monolayer on the Ir surface. Due to the lattice mismatch of Co and Ir, a substantial tensile strain of 8.4% in epitaxial Co films on Ir(001)-(5 × 1)-hex surfaces is expected. The nearest-neighbor distance of bulk fcc Ir and Co is a_{Ir} =2.715 Å and a_{Co} =2.507 Å, respectively.

Instead of the (5 × 1) symmetry, a new (1 × 1) symmetry takes place as Co is grown on the Ir(001)-(5 × 1)-hex surface. Figure 4.3 shows the observed (1 × 1) LEED pattern (100 eV) of 0.8 ML and 1.8 ML Co film on Ir(001)-(5 × 1)-hex, indicating the Co-induced structural modifications at the surface. Note that the superstructure LEED spots are weak and still visible.[1] It has been demonstrated by STM and IV-LEED studies [55, 57, 58] that upon around 0.1 ML Co coverage, the additionally accommodated Ir atoms in the hexagonal-close-packed arrangements are locally lifted out to the top-most layer in which they form Ir chains. The Ir chains are aligned parallel to the direction of the reconstruction lines (trenches) of the formerly reconstructed Ir surface.

It has been reported in literature [55, 57–59] that the transition-metal adatoms occupy in the empty sites between these Ir chains with the average lateral spacing of $5a_{Ir}$ (the spacing of individual Ir chains can be $3a_{Ir}$, $5a_{Ir}$ and $7a_{Ir}$ where a_{Ir} is the

---

[1]It has been reported that the IV-LEED intensity spectra are very similar for 0.8 and 1.8 ML Co films grown on hydrogen-free Ir(100)-(5 × 1)-hex and Ir(100)-(5 × 1)-H substrates [55,56]. This result indicates the local structures for these two cases are almost identical.

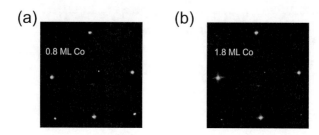

Figure 4.3: The observed (1×1) LEED patterns (100 eV) of (a) 0.8 ML and (b) 1.8 ML Co grown on Ir(001)-(5×1)-hex surface.

in-plane nearest-neighbor distance of Ir, $a_{Ir} = 2.715$Å [55, 58]). For thicker films, the modulation of the film arising from the presence of Ir chains at the interface has been demonstrated in Ref. [55, 57–59].

4.1.2 MOKE measurements

In order to investigate the magnetic anisotropy in more details for a 3 ML Co film grown on the reconstructed Ir(001), MOKE measurement are performed in three different geometries. (i) Magnetic field applied along the Ir$[1\bar{1}0]$ direction of a single (5×1) domain sample. (ii) Magnetic field applied along the Ir$[1\bar{1}0]$ direction of a multi-domain $(5 \times 1 + 1 \times 5)$ sample. (iii) Magnetic field applied along the Ir$[0\bar{1}0]$ direction of a single (5×1) domain sample. The relative orientation between magnetic field and the direction of the reconstructed lines of the Ir(001) substrate surface is shown in Fig. 4.4a-c. In Fig. 4.4a-c, the ball model of the Co films grown on Ir(001)-(5×1) is adapted from Ref. [55, 58, 60].

In bulk magnetic systems, the magnetocrystalline anisotropy energy corresponds to the differences in the magnetic energy when the sample is magnetized along two different principal crystallographic directions. These two directions are called easy and hard magnetization axis, respectively. As a result, the anisotropy energy barriers can effectively pin the magnetization along a certain direction in the bulk. Moreover, by lowering the sample dimension or breaking symmetry, such as a reduction of the coordination number or steps at the surface, may introduce an enhanced magnetocrystalline anisotropy.

In order to determine the direction of the easy magnetization axis, the results of the MOKE measurement of above cases (i-iii) are obtained on a 3 ML Co film grown on substrates having their long edge parallel to the Ir[110] or the Ir[100] direction (see Fig. 4.1a). The MOKE measurements are performed in the longitudinal geometry at room temperature. The angle of the incident and scattered laser beam with respect

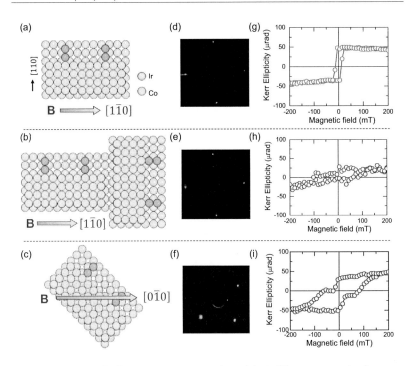

Figure 4.4: MOKE measurement are performed in 3 different geometries for a 3 ML Co film grown on reconstructed Ir(001). (a) Magnetic field applied along the Ir[1$\bar{1}$0] direction of a single (5 × 1) domain sample. (b) Magnetic field applied along the Ir[1$\bar{1}$0] direction of a multi-domain (5 × 1 + 1 × 5) sample. (c) Magnetic field applied along the Ir[0$\bar{1}$0] direction of a single (5 × 1) domain sample. The results of LEED (d-f) and Longitudinal MOKE (g-i) measurements are presented for the corresponding cases of (a-c). The ball model in (a-c) of the Co films grown on Ir(001)-(5 × 1) is adapted from Ref. [55, 58, 60].

to the film normal was 45°.

The MOKE hysteresis loops recorded on a 3 ML Co film on the reconstructed Ir(001) are presented in Fig. 4.4g-i. The results for the cases (i-iii) are illustrated as follows. For the case (i), the observed rectangular-like loop is a signature that the magnetic field is applied along an easy axis of magnetization. The result indicates the easy magnetization axis is perpendicular to the direction of the previously reconstructed lines of the substrate surface (see Fig. 4.4a).[2] For the case (ii), the orthogonal (5 × 1 + 1 × 5) reconstructed domains, the external magnetic field is applied simultaneously parallel and perpendicular to the orientation of Ir atomic chains. As shown in

---

[2]For indicating the orientation of the reconstructed Ir surface, in the following we refer to this direction as the Ir chains, as depicted in Fig. 4.4a-c.

Fig. 4.4h, the magnetization is not saturated under the maximum magnetic field (200 mT). In the case (iii) where the magnetic field is applied along the direction being 45° away from the long axis of Ir atomic chains, the loop shows two plateaus with two switching fields (±25 and ±63 mT) (Fig. 4.4c). The step-like MOKE hysteresis loop can be explained by the relative 45° orientation between the applied magnetic field and the easy axis in the sample plane. Thus, it is reasonable to assume that the hard-axis is parallel to the Ir atomic chains.

For a fourfold-symmetric Co(001) film on an atomically flat surface, a fourfold in-plane magnetic anisotropy is expected. For instance, in the case of Co/Cu(001), the easy magnetization axes lie in the Co$\langle110\rangle$ directions [61,62]. However, our results reveal that a large uniaxial magnetic anisotropy in the sample plane of the Co/Ir(001)-(5×1)-hex system is stabilized in the direction perpendicular to the Ir atomic chains.

Shape anisotropy and magnetocrystalline anisotropy are considered to be the two main sources for magnetic anisotropy. The easy magnetization axis in the system is determined by a competition between the shape anisotropy and the magnetocrystalline anisotropy. Shape anisotropy arises purely from dipolar interaction. For a given shape of a magnetic solid, the system tries to minimize the magnetic stray field and thereby its magnetostatic energy. Thus, for a Co film on Ir(001)-(5×1), the easy axis determined by the shape anisotropy parallel to the Ir atomic chains is expected. Therefore, the observed easy magnetization axis, being perpendicular to Ir atomic chains, cannot be accounted for by the shape anisotropy. We suggest the observed large uniaxial magnetic anisotropy in Co/Ir(001) is dominated by the magnetocrystalline anisotropy.

The anisotropic bonding environment near Ir chains at the interface may lead to remarkable consequences for the in-plane orbital moment anisotropy and a large in-plane uniaxial magnetocrystalline anisotropy. For instance, in the case of monatomic Co chains situated at both the step-edge of a Pt(111) and the (5×1) reconstructed Ir(001), the easy magnetization axis of the Co chains projected onto the plane was found to be perpendicular to Pt step edges and Ir atomic chains [63,64]. Nevertheless, for a contrast example of a Co film grown on vicinal Cu(001), the easy magnetization axis is aligned parallel to the step edges of the Cu substrate [62]. While the states of the Co film hybridize rather weakly with the closed $3d$-shell of Cu substrate in the region near the Fermi level, the unfilled and spatially more extended $5d$ states (Ir or Pt) can have a rather large energetic overlap with the $3d$ states of the Co film. The strong hybridization between $3d$ ultrathin films and $5d$ substrates may modify the electronic structures and result in bandwidth broadening and an unquenched (or an enhanced) orbital moment [65,66].

As noted by many studies, the interface-induced magnetocrystalline anisotropy energy is considerably sensitive to details of the atomic bonding and the changes in

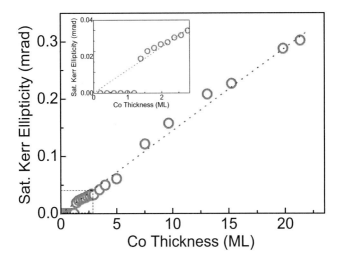

Figure 4.5: Kerr ellipticity in saturation (B = 200 mT) as a function of the Co thickness. The blue dotted line indicates a linear dependence of the Kerr ellipticity on the Co thickness. The inset shows a magnified part of the data. The magnetic loop is observed only when the thickness is above 1.4±0.2 ML, indicating the onset of ferromagnetic order at room temperature for Co/Ir(001).

the energy of the d orbitals of the interface Co atoms. The strength of the interface bonds has been successfully elucidated by an effective ligand interaction model [67]. Their results indicate that if the bonding strength of the same in-plane Co-Co bonding is normalized to 1.00, the ratio of the bonding strength at the interface, assuming the same in-plane Co-Co bonding as the out-of-plane Co-X bonding, is 0.83, 1.60 and 1.69 for $X =$ Cu, Pt and Ir, respectively. Correspondingly, for the X/Co/X multilayer system, the Co layer has a perpendicular anisotropy as it is sandwiched between $X =$ Ir or Pt layers, while the Co film with in-plane magnetic anisotropy occurs for $X =$ Cu (note that a freestanding Co monolayer with in-plane anisotropy has been predicted) [67–69]. Based on the effective ligand interaction model, the origin of the MCA can be qualitatively assessed in terms of the orbital moment anisotropy arising from the anisotropic bonding strength.

Our results provide a qualitative determination of the in-plane magnetic anisotropy in the ferromagnetic epitaxial Co/Ir(001)-(5 × 1) system by means of MOKE measurements. The easy magnetization axis in 3 ML Co/Ir(001) stabilized in the direction perpendicular to the Ir atomic chains is observed. A large in-plane uniaxial magnetic anisotropy in Co/Ir(001) dominated by the magnetocrystalline anisotropy, arising from the strong interfacial Co$_{3d}$-Ir$_{5d}$ electronic hybridization, is suggested.

We performed MOKE experiments as a function of Co thickness. The experiments are performed on single domain samples with the magnetic field applied along the Co[1$\bar{1}$0] direction. Fig. 4.5 presents the saturation Kerr ellipticity as a function of the Co thickness. The linear variation of Kerr ellipticity with Co thickness is observed at room temperature. In the thin-film limit, since the total thickness of the film is much less than the wavelength of the incident light and the interference effect between the individual layers is ignored, the Kerr effect is sensitive to the increasing amount of Co and obeys an additivity law. This law states the total Kerr signal is simply a summation of the Kerr signals from each individual magnetic layer [70,71]. One may determine an onset of ferromagnetism at room temperature, marked by the appearance of the hysteresis. The magnetic hysteresis loop is observed only when the thickness is above 1.4±0.2 ML, indicating the onset of ferromagnetic order at room temperature, which initially falls into the straight line as shown in the inset of Fig. 4.5. The rectangular-like hysteresis loop is observed for the film thicknesses lager than 1.6±0.2 ML.

4.1.3   SPEELS measurements

Magnons in Co/Ir(001)-(5×1) along the Co[110] direction

An example of the SPEELS spectra recorded at room temperature on a 3 ML Co film on Ir(001)-(5 × 1) at an in-plane wave-vector transfer of $\triangle K_\parallel = 0.7$ Å^{-1} is shown in Fig. 4.6a. The scattering plane is chosen to be parallel to the Co[110] direction, corresponding to the $\overline{\Gamma} - \overline{X}$ direction. The incident electron energy is 8 eV with an energy resolution of 23.8 meV. I_- and I_+ are referred to the measured intensities of scattered electrons when the spin polarization vector of the incoming electron beam is parallel and antiparallel to the Co[1$\bar{1}$0] direction, respectively, being the same as the direction of easy magnetization axis of the Co film. A pronounced peak at zero energy-loss ($E = 0$ meV) is found in both I_- and I_+ spectra, arising from the quasi-elastically scattered electrons during the scattering process (usually called quasi-elastic peak). Beside the quasi-elastic peak, one observes a well-resolved excitation peak, appearing at $\simeq 135$ meV in the energy loss region of the I_- spectrum. This result clearly demonstrates that only incident electrons with a polarization vector parallel to the magnetization direction (the minority spins) are allowed to excite magnons, while those with a spin antiparallel to the magnetization direction (the majority spins) are not allowed. Note that due to the electron spin being antiparallel to the magnetic moment in ferromagnets, the minority-character spin is referred to the sample magnetization direction [72]. The information regarding the energy and lifetime of the excited magnons can be obtained by analyzing the difference spectrum ($I_- - I_+$), represented by the green circles in Fig. 4.6.

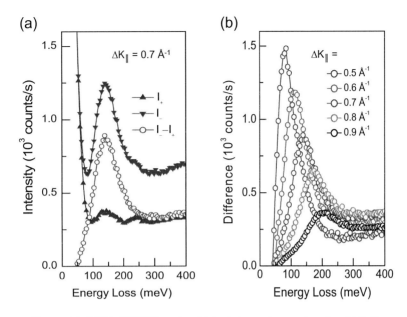

Figure 4.6: (a) The SPEELS spectra obtained at room temperature for a 3 ML Co film on a (5×1) reconstructed Ir(001) surface at an in-plane wave-vector transfer of $\triangle K_\parallel = 0.7$ Å^{-1} along the Co[110] direction. The intensity spectra I_+ (blue) and I_- (red) are denoted as the measured intensities of scattered electrons when the spin polarization of the incoming electron beam is parallel and antiparallel to the Co[$1\bar{1}0$] direction, respectively, being the same direction as the static magnetization of the Co films. The difference spectrum $(I_- - I_+)$ is shown by green circles. (b) A series of difference spectra recorded at various in-plane wave-vector transfers ranging from 0.5 to 0.9 Å^{-1} along Co[110] direction (corresponding to the $\overline{\Gamma} - \overline{X}$ direction in reciprocal space).

Series of difference spectra with various in-plane wave-vector transfers ranging from 0.5 to 0.9 Å^{-1} along the Co[110] direction (corresponding to the $\overline{\Gamma}-\overline{X}$ direction in the reciprocal space) are shown in Fig. 4.6b. As presented in Fig. 4.6b, the excitation energy increases as the wave-vector increases, exhibiting a clear energy–momentum dispersion. The amplitude of the magnon peak reduces as wave-vector increases. At the same time, the peak broadening increases. Owing to the itinerant nature of the Co films, strongly damped spectral features are expected as the spin excitations encounter the Stoner continuum of incoherent electron-hole excitations, especially for the spectra recorded at higher wave-vector transfers.

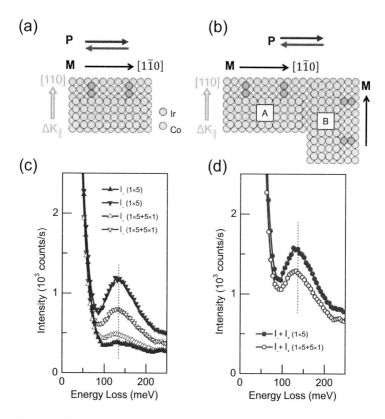

Figure 4.7: A schematic representation of a 3 ML Co film on Ir(001) with (a) (1×5) and (b) (1×5+5×1) surface reconstructions. The spin-polarization vector **P** of the incident electrons and the magnetization **M** is indicated. (c) The SPEELS I_- and I_+ spectra recorded for a 3 ML Co film on Ir(001) with (1×5) and (1×5+5×1) surface reconstructions at an in-plane wave-vector transfer of $\triangle K_\| = 0.7$ Å^{-1} along the Co[110] direction. The intensity spectra I_- (red) and I_+ (blue) are denoted as the measured intensities of scattered electrons when the spin polarization of the incoming electron beam is parallel and antiparallel to the Co[1$\bar{1}$0] direction. The total intensity spectra $(I_- + I_+)$ are displayed in (d). The vertical dashed-line is the guide to the eye. The spectra are recorded on the same Co film on Ir(001) with (5×1) and (1×5+5×1) reconstructed surfaces.

Magnons in Co/Ir(001)-(5 × 1 + 1 × 5) along the Co[110] direction

In order to investigate the role of in-plane magnetic anisotropy and the Ir surface reconstruction on the high-energy magnons, the same Co film grown on Ir(001)-(5 × 1 + 1 × 5) are investigated by SPEELS. In the case of Co films grown on the (5 × 1 + 1 × 5) reconstructed Ir surface, the in-plane wave vector of the excited magnons in the SPEELS measurements is along the Co[110] direction. This means that magnons propagate simultaneously parallel and perpendicular to the direction of the Ir atomic chains. Figures 4.7a and b show schematic representations of the relative orientation between the spin-polarization vector $\mathbf{P}$ of the incident electrons and the easy magnetization axis $\mathbf{M}$ for the case of Co films on (1×5) (Fig. 4.7a) and (1×5+5×1) (Fig. 4.7b) reconstructed Ir(001) surface.

Figure 4.7c shows the SPEELS spectra for a 3 ML Co film on Ir(001) with (1×5+5×1) surface reconstruction (open triangles) compared to the one on single (5×1) reconstructed surface (solid triangles). The data are recorded at an in-plane wave-vector transfer of $\triangle K_\parallel = 0.7$ Å^{-1}. The magnon peak at $\simeq 135$ meV is observed in both I_- and I_+ spectra. All spectra in Fig. 4.7c are recorded on the same Co film.

As discussed in Sec. 4.1.2, the in-plane easy magnetization is stabilized by a large in-plane uniaxial magnetic anisotropy for Co films grown on the (5 × 1) reconstructed Ir surface. The easy axes for Co films grown on Ir(001)-(5 × 1) and Ir(001)-(1 × 5) are mutually orthogonal to each other. Owing to the presence of the orthogonal domains, the inelastic excitation signals contain contributions from both (5×1) and (1×5) domains, denoted by A and B in Fig. 4.7b. Incoming electrons with spin polarization parallel to the magnetization at A domains can excite the magnons in this area while electrons with opposite spin are not allowed to do so. This fact leads to a peak in I_- spectra. In the case of B domains, the magnetization is orthogonal to the beam polarization. In such a case, one can show that a polarized electron beam acts as a fully unpolarized beam, composed of both majority and minority spins (a detailed information is given in Appendix A). It is therefore expected that the magnons should be observed in both I_- and I_+ spectra. This is exactly what we observe in Fig. 4.7c.

In other words, in the experiment, the magnons propagating parallel and perpendicular to the easy magnetization axis are probed simultaneously (i.e. $\triangle K_\parallel \| \mathbf{M}$ and $\triangle K_\parallel \perp \mathbf{M}$) (see Fig. 4.7b). Denoted by the dashed line in Fig. 4.7c, our results reveal that the energies of high wave-vector magnons do not depend on the static magnetization direction.

Comparing the SPEELS total intensity spectra of Co films grown on Ir(001)-(5 × 1) and Ir(001)-(5 × 1 + 1 × 5) indicates the nearly the same magnon energy (see Fig. 4.7d). This is a confirmation that the energies of the high wave-vector magnons are independent of the type of the surface reconstructions. We conclude that the high

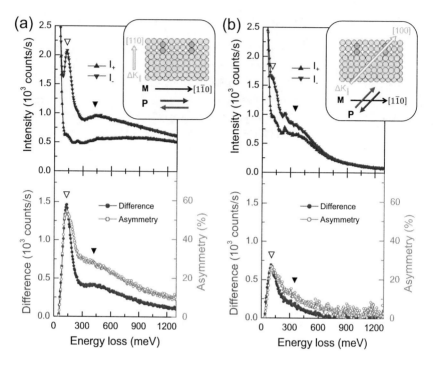

Figure 4.8: SPEELS spectra recorded on a 3 ML Co film grown on Ir(001) in the energy-loss range of $0-1300$ meV at an in-plane wavevector transfer of $\triangle K_\parallel = |0.7|$ Å^{-1}. The measurements are performed along (a) the Co[110] and (b) the Co[100] directions, corresponding to the high-symmetry directions $\bar{\Gamma} - \bar{X}$ and $\bar{\Gamma} - \bar{M}$ of the surface Brillouin zone, respectively. Two magnon peaks are clearly observed, indicated by the open and the solid black triangles. The beam polarization **P** of the incident electrons and the easy magnetization axis **M** are depicted in the insets.

wave-vector magnons are excited locally (on a length scale of a few nanometers), dominated by the short-range magnetic exchange interaction. Due to the fact that the magnetic anisotropy energy mainly arises from the much weaker spin-orbit or the dipolar interaction, its effect on the energies of high wave-vector magnons is negligible.

Magnons in Co/Ir(001)-(5×1) along the Co[100] direction

In the measurements reported in the last section, the magnon propagation direction was along the Co[110] direction. The results for the measurements along the Co[100] direction are explained in the following. As mentioned above, since the high wave-vector magnon energy does not depend on the type of the surface reconstructions, we show only the results of a Co film on the single domain (5×1) reconstructed Ir(001) surface. Fig. 4.8 shows the SPEELS spectra recorded in the energy-loss range of $0 - 1300$ meV for a 3 ML Co film grown on Ir(001) at an in-plane wavevector transfer of $\triangle K_\parallel = 0.7$ Å^{-1}. The measurements are performed along the Co[110] (Fig. 4.8a) and the Co[100] (Fig. 4.8b) directions, corresponding to the high-symmetry $\overline{\Gamma} - \overline{X}$ and $\overline{\Gamma} - \overline{M}$ directions of the surface Brillouin zone, respectively.

For the measurements performed along the Co[100] ($\overline{\Gamma} - \overline{M}$) direction, the incident beam polarization has a relative orientation of $\simeq 45°$ with respect to the sample magnetization (inset, Fig. 4.8b). As discussed in Sec. 4.1.3, assuming a fully polarized incoming electron beam, $\mathbf{P_0} = 1$, each individual beam of the opposite orientation carries different proportions of the minority and majority electrons (see Appendix A). The observed different spectral weight in I_+ and I_- spectra reflects that the two opposite-oriented incident beams have different amounts of electrons with minority spin, due to the fact that only the incident electrons of minority spin character can excite magnons via the exchange scattering process. Thus, the largely reduced intensity of the magnon peaks in the difference $(I_- - I_+$, green) and the asymmetry $(\frac{I_- - I_+}{I_- + I_+}$, orange) spectra is observed as shown in Fig. 4.8b, compared to the ones in Fig. 4.8a. Our results demonstrate that the spectral amplitude in SPEELS spectra strongly depends on the relative orientation between the incident beam polarization and the sample magnetization.

Interestingly, several magnon peaks are clearly observed for spectra recorded along both high-symmetry directions. The resolved high loss energy peak appears as a broad shoulder (the solid black triangles) beside a rather sharp low energy peak (the open black triangles), which is attributed to the standing spin-wave modes. The detailed interpretations of all magnon modes in the 3 ML Co/Ir(001) system are discussed in the next subsection.

Direct observation of the confined magnon modes

For the case of a three-atomic-layer ($N = 3$) ferromagnetic film, the "wave-vector" of standing spin waves confined perpendicular to the film surface, resulting from the quantization of the continuous bulk magnon branch, is given by $q_\perp \simeq \frac{n\pi}{d}$, where d is the thickness of the film and $n = 0$, 1 and 2 refers to the quantum number of the confined magnon modes, as discussed in Sec. 2.1.3.

Figure 4.9 shows series of difference and total intensity spectra measured on a 3 ML Co film grown on Ir(001) for wave vector transfer along the Co[110] ($\overline{\Gamma} - \overline{X}$) (Fig. 4.9a) and the Co[100] ($\overline{\Gamma} - \overline{M}$) (Fig. 4.9b) directions. From top to bottom, the in-plane momentum transfer varies from near the center to the boundary of the surface Brillouin zone. To point out the dispersive behavior of each magnon mode, the low-

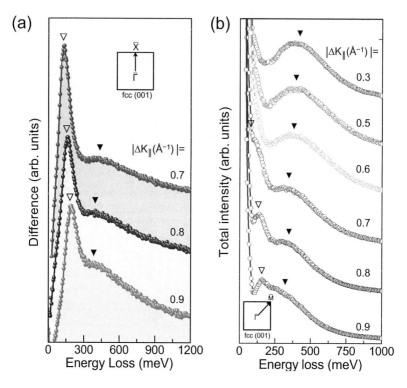

Figure 4.9: The wave-vector dependence of the SPEELS (a) difference and (b) total spectra for a 3 ML Co/Ir(001). The measurements are performed along (a) the Co[110] ($\overline{\Gamma}-\overline{X}$) and (b) the Co[100] ($\overline{\Gamma}-\overline{M}$) directions (shown in the inset). The open and solid black triangles indicate the peak positions of the two overlapping modes with $n = 0$ and $n = 1$ (open triangle) and the high-energy mode with $n = 2$ (solid triangle).

($n = 0$ and 1) and the highest-energy ($n = 2$) excitation peaks are represented by the open and solid black triangles, respectively. One can observe that the peak position significantly changes with an increase of the in-plane wave-vector. The distinct wave-vector dependent behavior of the two peaks are clearly demonstrated. Note that the excitation energy is independent (within error bars) of the sign of the in-plane wave-vector transfer in the Co/Ir(001) system. Thus, the absolute value of in-plane wave-vector transfer is shown in Fig. 4.9.

For the magnons investigated here (the exchange-dominated magnons), the full-width at half-maximum (FWHM) of the magnon peak is comparable to the magnon energy. Especially at a larger wave-vector transfer ($\triangle K_{\parallel} \geq 0.7$ Å^{-1}), the energy difference between two low-energy modes is much smaller than the intrinsic peakwidth of each particular magnon mode. Thus, the hardly distinguishable peaks show up as a single spectral profile in SPEELS spectra. For the spectra measured along the Co[110] and the Co[100] directions shown in Fig. 4.8 and Fig. 4.9, the two-peak profile denoted by the open and solid black triangles is identified as the the peak positions of the two overlapping modes with $n = 0$ and $n = 1$ (open triangle) and the high-energy mode with $n = 2$ (solid triangle).

In measurements along both the $\overline{\Gamma} - \overline{X}$ and the $\overline{\Gamma} - \overline{M}$ directions (Fig. 4.9), the sharp low-energy mode shifts to higher energies as the in-plane wave-vector transfer increases, while the well-defined highest-energy mode disperses toward lower energies. The energy difference between the low and highest energy modes decreases with the wave-vector increasing. The reduced energy difference between low- and highest-energy modes indicates the opposite sign of the dispersion of the low- and highest-energy modes.

Magnon dispersion relation of confined magnon modes

In order to obtain the magnon dispersion relation and the lifetime of all magnon modes, the difference spectra $(I_- - I_+)$ are fitted by three Lorenzian lineshapes. Figure 4.10 presents systematically analyzed fitting results of a series of difference spectra at in-plane wave-vector transfers ranging from 0.3 to 0.6 Å^{-1} along the Co[100] $(\overline{\Gamma} - \overline{M})$ direction. The blue, green and pink curves correspond to the confined magnon modes with the quantum number of $n = 0$, 1 and 2, reflecting the characteristic features of standing spin waves with zero, one and two node(s) inside the film.

In Fig. 4.10, well-separated three magnon modes at small in-plane wave-vector transfer are observed. As one can see, the two low-energy modes begin to merge at $\triangle K_\parallel = 0.5$ Å^{-1} and all three peaks become moderately closer when approaching the zone boundary. The spectral line shapes of the two low-energy peaks become weaker in intensity and broader in energy with an increase of the in-plane wave-vector transfer, while the opposite tendency is observed for the highest-energy mode. The peak-width of the three Lorentzians is related to the intrinsic damping of each magnon mode. The lowest-energy excitation peak $(n = 0)$ exhibits a high and narrow spectral feature, indicating the long-lived coherent spin-precessing mode. The less sharp profile is found for the second-lowest energy mode $(n = 1)$. For the highest-energy mode $(n = 2)$, the peakwidth becomes substantially broad, even at a small in-plane wave-vector transfer $(\triangle K_\parallel < 0.3$ $\text{Å}^{-1})$, indicating a relatively short lifetime of this mode.

By plotting the energy as a function of the in-plane wave-vector transfers, we obtain the dispersion relation of all three modes, as shown in Fig. 4.11. In contrast to the "upward" dispersion relation of the two low-energy magnon branches, the highest-

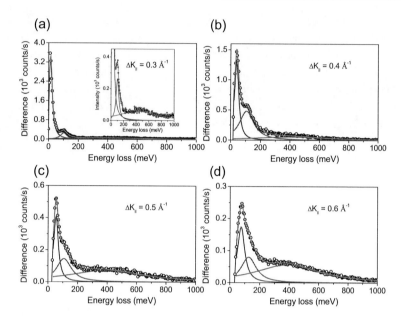

Figure 4.10: Series of difference spectra $(I_- - I_+)$ recorded on 3 ML Co/Ir(001)-(5×1) in the energy-loss range of $0 - 1$ eV. The in-plane wave-vector transfer is varied from 0.3 to 0.6 Å^{-1}. The wave-vector of magnons was probed along the Co[100] direction. The open black circles represent the experimental data. The blue, green and pink lines correspond to the confined magnon modes with the quantum number of $n = 0$, 1 and 2, respectively. The orange line shows the sum of the three Lorentzians. The inset in (a) shows a magnified part of the data.

energy magnon mode shows a "downward" dispersion relation with the negative group velocity $[v_g = \partial_{K_\parallel} E(K_\parallel)]$.

The intrinsic broadening of the high-energy magnon modes suggests a relatively short lifetime. This result implies that, even at a small in-plane wave-vector transfer, the standing spin waves with a large confined "wave-vector" along the surface normal suffer from a large Landau damping. For the first time, the well-defined confined highest-energy magnon mode in a 3 ML tetragonal Co film is experimentally observed and its peculiar dispersion relation is revealed.

In summary, the in-plane-momentum-resolved confined magnon modes are investigated by SPEELS. In the 3 ML Co/Ir(001)-(5×1) system, our results provide the experimental evidence of the distinctive dispersive behavior of each magnon mode. These aspects will be discussed in Sec. 5.2 and 5.5.

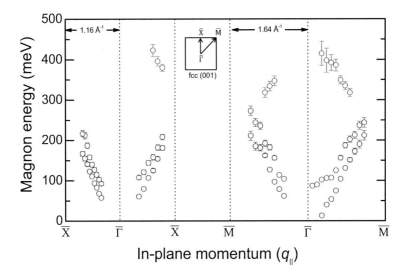

Figure 4.11: The measured magnon dispersion relation of a 3 ML Co film grown on Ir(001)-(5×1) obtained from the peak positions in Fig. 4.10. The confined magnon modes corresponding to the quantum number of $n = 0$ (blue), $n = 1$ (green) and $n = 2$ (pink) are presented. The length of the $\overline{\Gamma} - \overline{X}$ is 1.16 Å^{-1} and that of $\overline{\Gamma} - \overline{M}$ is 1.64 Å^{-1}. The in-plane nearest-neighbor lattice constant of a Co film is assumed to be the same as the Ir substrate.

4.2 The Co/Cu(001) system

It is well-known that the Cu(001) surface is an ideal template for growing ultra-thin Co(001) films due to the small lattice mismatch (being less than 2%). Here we report on the experimental results of high-energy magnetic excitations in a 3 ML face-centered tetragonal (fct) Co film grown on Cu(001) surface and compare the results to the ones of the 3 ML Co/Ir(001)-(5×1) system. Details regarding sample preparation and characterization are reported in [73].

4.2.1 SPEELS measurements

Figure 4.12 shows SPEELS difference spectra obtained on a 3 ML Co film on Cu(001) with various in-plane wave-vectors ranging from -0.3 to -0.7 Å^{-1} along the Co[110] direction ($\overline{\Gamma} - \overline{X}$ in reciprocal space). The SPEELS difference spectra are fitted using three Lorentzians, as described in Sec. 4.1.3. The resulting dispersion relation of all three magnon modes is presented in Fig. 4.13. As seen in Fig. 4.13, the highest magnon mode shows almost no dispersion, while the two low-energy magnon modes show the

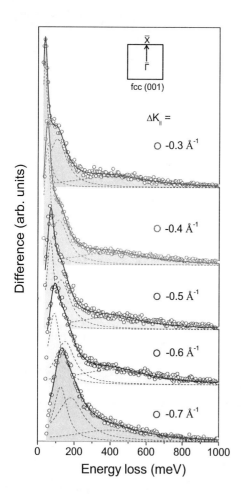

Figure 4.12: Series of SPEELS difference spectra recorded on 3ML Co/Cu(001) with in-plane wave-vectors $\Delta K_\parallel$ ranging from -0.3 to -0.7 Å^{-1} along the Co[110] direction ($\overline{\Gamma} - \overline{X}$). The spectra are obtained with a primary beam energy of $E_0 = 8$ eV. The experimental data (open circles) are well described by the fits (dashed lines) of the three Lorenzian lineshapes, corresponding to the confined magnon modes of $n = 0, 1, 2$. The solid line shows the sum of the three Lorenzians.

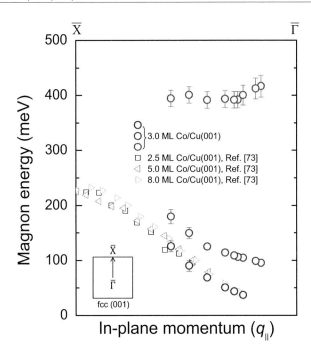

Figure 4.13: The measured magnon dispersion relation of 3 ML Co/Cu(001) ($\bigcirc$) along the Co[110] ($\overline{\Gamma} - \overline{X}$) direction, compared with the one of a 2.5 ML ($\Box$), a 5 ML ($\triangleleft$), and a 8 ML ($\triangleright$) Co films on Cu(001) adapted from Ref. [73]. The confined magnon modes with the quantum number of $n = 0$ (blue), $n = 1$ (green) and $n = 2$ (pink) are presented. The length of the $\overline{\Gamma} - \overline{X}$ direction is 1.23 Å^{-1}.

typical "parabolic" dispersion relation. These two low energy modes, corresponding to the standing spin waves with zero- and one node(s), show an energy separation of about 70 meV near the zone center ($\overline{\Gamma}-$ point), while it is much smaller than the one of the magnon modes of $n = 1$ and $n = 2$, being 300 meV. The similar energy interval among distinct magnon modes near the zone center ($\triangle K_{\parallel} = 0.3$ Å^{-1}) was also found in the Co/Ir(001) system.

The measured intensity (cross section) for each magnon mode in SPEELS strongly depends on the kinetic energy of the incoming electrons E_i in the same scattering geometry. Figure 4.14 shows the difference spectra recorded at an in-plane wave-vector transfer of -0.4 Å^{-1} with the incident electron energies of 5 eV and 8 eV. One clearly observes that the relative spectral weight between the two magnon modes in the SPEELS difference spectrum is extremely sensitive to the impact energy. Remarkably, the high-energy peak, appearing as a shoulder, is more pronounced at the higher

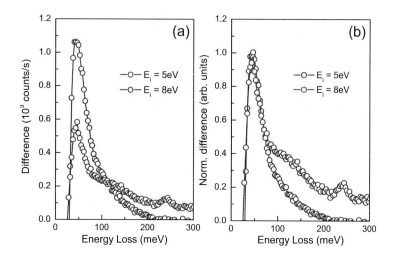

Figure 4.14: (a) The SPEELS difference spectra recorded on an in-plane wavevector transfer of -0.4 Å^{-1} along the Co[110] direction with the incident electron energy of 5 eV (blue circles) and 8 eV (red circles). The normalized spectra are shown in (b). The normalization is done with respect to the intensity of the quasi-elastic peak in the total $(I_- + I_+)$ spectrum. The highest normalized intensity is set to 1. The small and sharp peak at about 240 meV in spectra is attributed to the vibrational excitations peak of a small-amount adsorbed CO molecules on the sample surface.

impact energy, whereas the intensity of the low energy peak is larger at the lower impact energy.[3] The dependence of the peak intensity on the impact energy for each particular magnon mode is also observed in the Co/Ir(001) system. Thus, by fine-tuning the incident electron energy, one can change the relative intensity of each magnon mode, making the SPEELS sensitive to the desired mode.

---

[3]In Fig. 4.14, besides the magnon peak, a minor peak is found at $\simeq 240$ meV, which is attributed to the vibrational excitations arsing from small-amount adsorbed CO molecules on the sample surface. However, they are much below the detection limit of Auger spectroscopy. Note that the cross section of the surface vibrational excitations is roughly three orders in magnitude larger than the one of spin excitations in ferromagnets [74]. The observed same magnon peak position, compared to the one measured in a clean sample, suggests the magnon energy is unaffected by small-amount vibrational excitations.

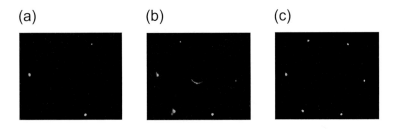

Figure 4.15: Typical LEED patterns of (a) a clean Pt(111) substrate, (b) 3 ML Co/Pt(111) and (c) 8 ML Co/Pt(111) taken at a primary beam energy of 75 eV.

4.3 The Co/Pt(111) system

4.3.1 Sample preparation and characterization

Before deposition of cobalt, a clean and well-ordered Pt(111) surface was prepared by multiple cycles of Argon ion sputtering (1.5 keV) and subsequently annealing at about 900 K in a base pressure of 3×10^{-11} mbar. The chemical cleanness and surface roughness of the substrates were verified by LEED, AES and EELS. As described in Sec. 4.1.1, the film is grown by electron-beam evaporation from a high-purity Co rod. The film thickness and the evaporation rate are quantitatively calibrated by means of MEED intensity oscillations recorded during the evaporation, and are cross-checked by the Kerr signal in LMOKE measurements.

Figure 4.15a-c shows typical LEED patterns with the primary electron energy of 75 eV taken on the clean Pt(111) substrate, 3 ML Co/Pt(111), and 8 ML Co/Pt(111) at room temperature. The LEED pattern of the clean Pt(111) surface clearly shows threefold symmetry, corresponding to the fcc (111) surface. At low coverage of Co ($\leq$ 3 ML), the LEED pattern gradually evolves into the six characteristic satellite spots surrounding each integer spots. In agreement with the STM studies [75], the "unrotated" satellite spots reflect that the crystallographic axes of the Co film are oriented with the substrate. At higher coverages, the satellites are nearly invisible. Instead, the LEED pattern transforms into a six-fold symmetry which mainly contributes from the Co layers (Fig. 4.15c).

For the LEED pattern taken for low coverage of Co, the observed hexagonal-symmetric satellite spots arise from the Moiré lattice formed by the Co layers on the Pt(111) substrate. This imperfect layer-by-layer growth arises from the considerable lattice mismatch between Co and Pt, being about 10.4% (the nearest-neighbor distance of bulk hcp Co and fcc Pt is 2.51 Å and 2.77 Å, respectively). It should be noted that the Moiré pattern arises because the Co layers are relaxed (i.e., they do not have the same in-plane lattice constant as the substrate, in contrast to epitaxial growth). STM studies have shown a persisting incommensurate structure of the Moiré pattern up to 5 ML Co [75–77]. Moreover, those studies have shown that the cell size of the Moiré pattern of the Co layers in real space agrees well with the distance between an integer diffraction spot and its surrounding satellites in reciprocal space of the LEED pattern. By comparing the periodicity of the Moiré structure of the Co film with the in-plane lattice constant of Pt(111), the in-plane lattice constant of Co layers is contracted by $(7.9 \pm 0.7)\%$ with respect to the one of the Pt(111).

The satellite spots in the LEED pattern become nearly invisible when the Co thickness is larger than 5 ML, which is consistent with the persisting Moiré superstructures up to 5 ML [77, 78]. It has been reported in Ref. [77, 78] that the quasi-layer-by-layer growth becomes a 3D growth mode as the Co thickness is larger than 5 ML. Moreover, for a film thickness above 8 ML, the coexistence of oppositely-oriented triangle-shaped islands due to the reversed fcc stacking sequences of Co islands has been reported in Ref. [77, 78]. This finding explains the observed six-fold-symmetry in the LEED pattern for 8 ML Co/Pt(111) (Fig. 4.15c). The structural findings illustrated above provide important insights into the understanding of the magnetic anisotropy in the Co/Pt(111) system, as will be discussed below.

4.3.2   MOKE measurements

The magnetic properties of the Co films grown on Pt(111) are investigated by LMOKE. Figure 4.16 presents the saturated Kerr ellipticity as a function of the Co thickness at room temperature. Below 4 ML, zero Kerr ellipticity is found within the range of the applied field (± 200 mT). It has been observed in Refs. [79, 80] that 1 ML Co/Pt(111) has an easy magnetization direction perpendicular to the surface with the coercive field of 24.5 ± 1.5 mT. Co films show a spin reorientation transition (SRT) from out-of-plane to in-plane at around 4 ML [81–84]. The observed zero Kerr ellipticity for films with thicknesses below 4 ML is due to the fact that the magnetization is oriented perpendicular to the film. As will be demonstrated in Sec. 4.3.3, SPEELS experiments show that for films with thicknesses below 4 ML exhibit ferromagnetic order and the easy magnetization axis is found to be out-of-plane direction. Above 5 ML, Kerr ellipticity increases linearly with the film thickness, denoted by the dotted straight line extrapolated to the zero-thickness in Fig. 4.16. The observed rectangular hysteresis loop (inset of Fig. 4.16) indicates that the Co films have an in-plane magnetic anisotropy for film thicknesses above 5 ML.

The two dominant regimes shown in Fig. 4.16 are distinguished by the different magnetic anisotropy of the Co films. The spin-reorientation transition (SRT) from perpendicular to in-plane magnetization in Co/Pt(111) system occurs in the film

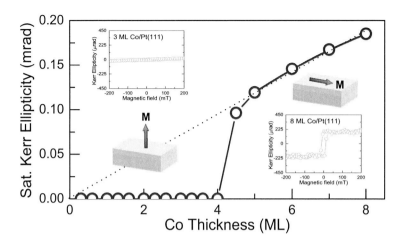

Figure 4.16: The saturated Kerr ellipticity as a function of the Co thickness, as measured by LMOKE. The magnetic field is applied along the $[11\bar{2}]$ direction of the Pt substrate. Below 4 ML, the zero Kerr ellipticity is found within the range of the applied field (± 200 mT). The spin-reorientation transition from perpendicular to in-plane magnetization occurs in the film thickness between 4 and 5 ML. Above 5 ML, Kerr ellipticity increases linearly with the film thickness, denoted by the dotted straight line extrapolated to the zero-thickness. The hysteresis loops of 3 ML and 8 ML Co films are shown in the inset.

thickness between 4 and 5 ML. Consistently, in the same thickness regime $(4-5$ ML), the transition of the growth mode from quasi-2D to 3D growth mode of the Co films grown on Pt(111) also takes place as reported in Refs. [76, 77, 81].

A large perpendicular magnetic anisotropy in Co films driven by the induced interface magnetocrystalline anisotropy arising from the strong interfacial Co_{3d}-Pt_{5d} electronic hybridization is suggested [65]. For thicker films, the shape anisotropy (dipole interaction) dominates over the interface magnetocrystalline anisotropy. Thus, with increasing film thickness an in-plane magnetic anisotropy is energetically preferred.

4.3.3 SPEELS measurements

Magnons in Co/Pt(111) with out-of-plane/in-plane magnetic anisotropy

Figure 4.17 shows the SPEELS spectra recorded on 3 and 8 ML Co films grown on Pt(111) at an in-plane wave-vector transfer of $\triangle K_\parallel = 0.8$ Å^{-1}. In the SPEELS measurement, the scattering plane is chosen to be parallel to the $[1\bar{1}0]$ $(\overline{\Gamma} - \overline{K})$ direction. The polarization vector of the incident beam is parallel and antiparallel to the $[11\bar{2}]$ direction. For 8 ML Co, the film is magnetized along the $[11\bar{2}]$ direction before the

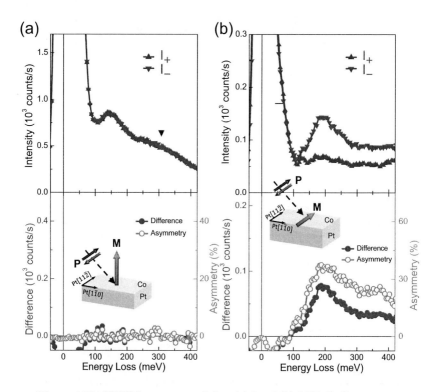

Figure 4.17: SPEELS spectra recorded on (a) 3 and (b) 8 ML Co films grown on Pt(111) at an in-plane wave-vector transfer of $\triangle K_\parallel = 0.8$ Å^{-1} probing along the $\overline{\Gamma} - \overline{K}$ ([1$\overline{1}$0]) direction of the surface Brillouin zone at room temperature. The incident electron energy is 4 eV. The solid black triangle in (a) indicates the confined magnon mode with the quantum number of $n = 2$ (The result of a fit for three magnon modes is shown in Fig. 4.18). For 8 ML Co, the film is magnetized along the [11$\overline{2}$] direction before the SPEELS measurement.

SPEELS measurement. The SPEELS I_- and I_+ spectra are referred to the measured intensities of scattered electrons when the spin polarization vector of the incoming electron beam is parallel (the red arrow) and antiparallel (the blue arrow) to the $[11\overline{2}]$ direction, depicted in the inset of Fig. 4.17.

For 3 ML Co with perpendicular magnetization, the polarization vector $\mathbf{P}$ is perpendicular to $\mathbf{M}$. As discussed in Sec. 4.1.3, in such a case, the magnon peak can be observed in both I_+ and I_- spectra. We attribute the resolved high energy-loss peak at $E_{loss} \simeq 305$ meV to the confined magnon mode with the quantum number of $n = 2$, denoted by the solid triangle in Fig. 4.17a. This confined magnon modes will be discussed in detail in the following section. However, for an 8 ML Co film with in-plane magnetic anisotropy, the polarization vector $\mathbf{P}$ of the incident beams is parallel and antiparallel to the magnetization direction $\mathbf{M}$. We observe a pronounced peak only in the I_- spectrum (Fig. 4.17b).

We conclude that, for 8 ML Co/Pt(111), the amplitude of magnon peaks is maximum when the polarization vector of the incident electron is parallel to the sample magnetization direction in the I_- spectrum (Fig. 4.17b), whereas the minimum intensity takes place when it is antiparallel to it (the I_+ spectrum in Fig. 4.17b). In such a situation, the maximum magnitude of the difference intensity and the asymmetry is obtained. For a contrast example of the 3 ML Co/Pt(111) system with perpendicular magnetic anisotropy, no asymmetry and difference is found as the oppositely-polarized beams are perpendicular to the sample magnetization direction.

Direct observation of the confined magnon modes

Figure 4.18 presents a series of the normalized total and the difference SPEELS intensity spectra recorded on 3 ML and 8 ML Co films on Pt(111). The spectra are recorded at different wave-vectors along the $\overline{\Gamma} - \overline{K}$ direction. From top to bottom in Fig. 4.18, the in-plane wave-vector transfer varies from near the center to near the boundary of the surface Brillouin zone. As described in Sec. 4.1.3, the SPEELS spectra for both 3 and 8 ML Co/Pt(111) are fitted using Lorentzians. The results of such analysis are shown in Fig. 4.18. A dependence of the magnon peaks for each magnon mode on the in-plane momentum transfer is observed.

For the case of the 3 ML sample, the well-resolved spectral feature of the highest-energy confined magnon mode with the quantum number of $n = 2$, marked by the black triangles in Fig. 4.18a, is clearly observed up to $\triangle K_\parallel = 0.8$ Å^{-1}. For the two low-energy modes ($n = 0, 1$), we observe an energy increases with wave-vector, exhibiting the typical "parabolic" magnon dispersion relation. However, in the case of the highest-energy mode ($n = 2$), a dispersion-less feature is observed.

For the case of the 8 ML sample, three magnon peaks show a clear dispersion. The

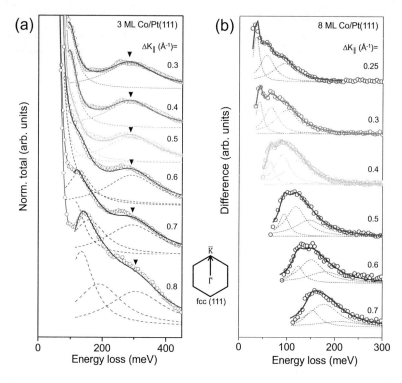

Figure 4.18: Series of SPEELS spectra for (a) 3 ML and (b) 8 ML Co/Pt(111) with various wave-vector transfers along the $\overline{\Gamma} - \overline{K}$ direction. The respective normalized total intensity and the difference intensity for 3 and 8 MLs are shown. The normalization factor in (a) is the intensity of the quasi-elastic peak in the total $(I_- + I_+)$ spectrum. The solid line shows the sum of the Lorenzian peaks (dashed lines). The experimental data (open circles) is well represented by the fits of the Lorenzian lineshapes. The well-resolved spectral feature for the highest-energy confined magnon mode ($n = 2$) in (a) is clearly visible up to $\triangle K_\parallel = 0.8$ Å^{-1}, marked by the black triangles. The three Lorenzian lineshapes in (b) from low- to high-energy peak positions at each wave-vector indicate the magnon modes with the quantum number of $n = 0, 1$ and 2, respectively.

energy position of the three magnon modes with the quantum number of $n = 0, 1$ and 2 increases with the wave-vector. Differing from the case of 3 ML, the magnon peak of the confined magnon mode of $n = 2$ is barely resolved, particularly for $\triangle K_{\parallel} > 0.5$ Å^{-1}. Although based on the Heisenberg model eight magnon modes are expected for an 8 ML film, the higher-energy magnon modes $(n > 2)$ are heavily damped and are hardly seen in the measured spectra. Therefore, the dramatically suppressed spectral feature of the higher-energy modes makes it practical difficult to capture the characteristics of all magnon modes in thicker film, even at a small in-plane wave-vector transfer.

For both film thicknesses, the energy separation among different magnon modes decreases with an increase of the wave-vector. By comparing the energies of the confined magnon modes with the same quantum number $n = 1$ or 2 between a 3 ML and an 8 ML, the much lower excitation energy in the 8 ML film is observed, which arises from the standing spin waves with the smaller "quantized wave-vector" perpendicular to the film surface. Note that at $q_{\parallel} = 0$ the approximate energies of the nth confined magnon modes are $D(q_{\perp}^n)^2 = D(\frac{n\pi}{d})^2$ where d is the thickness of the film. As a consequence, for the wave-vector transfer larger than $\triangle K_{\parallel} = 0.7$ Å^{-1} in an 8 ML Co film, the observed magnon broad spectral feature reflects the energy separation between the two magnon modes is much smaller than their intrinsic peakwidth. The large intrinsic magnon peak broadening at the higher wave-vector makes the identification of different magnon modes very difficault.

Figures 4.18a and b clearly show at a certain wave-vector, the broadening in peakwidth with an increase of the quantum number of the confined magnon modes. This means that the magnon lifetime becomes shorter as the quantum number of magnon modes increases. In the case of the 8 ML sample, the same confined magnon modes of $n = 1$ or $n = 2$ possess a smaller excitation energy and narrower peakwidths. This is due to the fact that for this case the magnon "quantized wave-vector" perpendicular to the film surface corresponding to $n = 1$ and 2 modes is smaller.

It has been established by theoretical calculations [15, 85] that in itinerant ferromagnets the spectral peakwidth of the magnetic excitations is dominated by the competition between the formation of magnons, described by well-defined atomic magnetic moments, and the attenuation from the single-particle Stoner excitations. The degree of the attenuation of magnons depends on the density of Stoner states near the Fermi level in the respective energy-momentum space of magnons. The Stoner density of states depend strongly on the details of the electronic structures.

The rather broad spectral feature for the higher-energy mode arises from the hybridization of the magnons with the Stoner-excitation. Moreover, for 8 ML, even at a small in-plane wave-vector ($\triangle K_{\parallel} = 0.3$ Å^{-1}), the absence of the confined magnon modes with the quantum number of $n > 2$ implies that they are strongly damped. For wave-vectors larger than $\triangle K_{\parallel} > 0.7$ Å^{-1}, the damping becomes more severe so

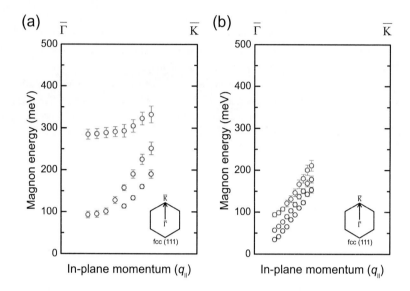

Figure 4.19: The measured magnon dispersion relation of (a) a 3 ML Co and (b) an 8 ML Co on Pt(111) along the $\overline{\Gamma} - \overline{K}$ direction. The confined magnon modes with the quantum number of $n = 0$ (blue), $n = 1$ (green) and $n = 2$ (pink) are presented. The length of the $\overline{\Gamma} - \overline{K}$ direction for 3 ML and 8 ML films is 1.51 and 1.66 Å^{-1}, corresponding to the nearest-neighbor lattice constant of the Pt substrate and the bulk hcp Co, respectively.

that only a single broad feature is observed, instead of several resolvable peaks.

In summary, our results experimentally show the characteristics of the three magnon modes with quantum numbers of $n = 0, 1$ and 2 in 3 ML and 8 ML Co films grown on Pt(111). The damping of the magnon modes increases significantly with the energy and wave-vector.

Magnon dispersion relation of confined magnon modes

As mentioned above, the SPEELS spectra for 3 ML and 8 ML Co/Pt(111) are fitted using Lorentzians (Fig. 4.18). The energy position of each magnon modes as a function of wave-vector is directly obtained. The resulting dispersion relation of the three magnon modes is presented in Figs. 4.19a and b. In the case of the 3 ML sample, the two low-energy magnon modes show typical "parabolic" dispersion relation, while the less dispersive feature for the highest energy magnon mode is observed. In more detail, for the highest-energy magnon branch, up to around the middle of the Brillouin zone, the dispersion tends to be flat, i.e., no dependence of the magnon energy on wave-vector is observed. Then, with approaching the zone boundary, it gradually moves toward higher energy. Notably, for 3 ML Co/Pt(111), the dispersion relation differs markedly in comparison with the one in 3 ML Co/Ir(001), discussed in Sec. 4.1.3, and 3 ML Co/Cu(001) (Sec. 4.2.1). Moreover, the greatly lowered magnon energy for the magnon modes with $n = 2$ is observed in the case of 3 ML Co/Pt(111).

For the wave-vector of $\triangle K_{\parallel} = 0.3\ \text{Å}^{-1}$, the energies of the confined magnon modes with $n = 1$ and $n = 2$ for 3 ML films are around 98 and 287 meV and the corresponding ones for 8 ML films are around 66 and 98 meV, respectively. These two confined magnon modes with $n = 1$ (green circles in Fig. 4.19) and $n = 2$ (pink circles in Fig. 4.19) correspond to the standing spin waves with one and two nodes, respectively. We find that the magnon energy increases significantly as the thickness of the film decreases, especially for the confined magnon mode with $n = 2$. The considerably large energy separation between these two magnon modes of about 189 meV for a 3 ML is obtained, comparing to the one of about 32 meV for an 8 ML. However, with increasing wave-vector, the size of the energy separation between the confined magnon modes of order $n = 1$ and $n = 2$ is strongly reduced for a 3 ML. Our results clearly demonstrate the characteristics of the confined magnon modes are strongly thickness-dependent.

We conclude by noting that although the magnetic excitations for Co and other $3d$ transition metal ferromagnets have been studied for many decades, for the first time, we directly observe the spatially highly confined magnon modes in ultrathin Co films and their dispersion relation over a large part of the surface Brillouin zone. In contrast to the rather robust two low-energy modes, the peculiar dispersion relation of the highest-energy mode for 3 ML Co films grown on Ir(001), Cu(001) and Pt(111) substrates are experimentally revealed. The novel observation of the highly confined magnon mode in ultrathin Co films provides the essential information to understand the thermodynamic, transport, excitation and magnetic properties in the ultrathin magnetic films.

Chapter 5

Discussion

The aim of this work is to investigate the magnetic excitations in ultrathin ferromagnetic Co films grown on different substrates. The focus is put on the characteristics of magnon modes with the quantum number of $n = 0, 1$ and 2, as discussed in Chapter 4. For a quantitative understanding of the results presented above, a simple Heisenberg model is applied to understand the dispersion relation of all confined magnon modes. Additionally, first-principles adiabatic spin dynamics calculations are used in order to address the role of the substrate, film structure and electronic band structure on the magnon dispersion relation and the interatomic magnetic exchange interaction. The first-principles calculations were performed by Arthur Ernst at the Max-Planck-Institute in Halle, and are discussed here in the context of the experimental work. A quantitative determination of the lifetime of the confined magnon modes is provided as well.

5.1   Film structure and substrate dependence

In this section we aim to compare the magnon dispersion relation measured for different systems. Before we discuss the characteristics of different magnon modes of each system, let us briefly discuss the structural and electronic properties of each system.

5.1.1   "Acoustic" magnons in Co and Fe films

Table 5.1 summarizes the electronic configuration and the structural parameters for Cu, Ir, Pt, Rh and W. The epitaxial misfit (η) and the easy magnetization axis for the pseudomorphic growth of the Co films on different substrates are indicated. In this section, three different kinds of the substrate surface orientations are presented and compared, i.e., fcc(001) (Cu, Ir and Rh), fcc(111) (Pt) and bcc(110) (W) substrate surfaces. Note that the magnetic excitations for the 3 ML Co film grown on Rh(001) were also studied in the course of this thesis. The reported results in the literature of

Table 5.1: The electronic configuration and structural parameters of Cu, Ir, Pt, Rh and W. The epitaxial misfit (η) is defined as $\eta = \frac{a_s - a_f}{a_f}$, where $a_{s(f)}$ is the in-plane atomic spacing of the substrate (film), assuming the pseudomorphic growth of Co films on different substrates. The in-plane ($\parallel$) or out-of-plane ($\perp$) easy magnetization direction for a 3 ML Co on each substrate is indicated.

substrate	Valence electrons	crystal structure	in-plane lattice constant, a (Å)	epitaxial misfit, η (%)	easy axis for a 3 ML Co film
Cu	$3d^{10}4s^1$	fcc	3.61	1.7	$\parallel$
Ir	$5d^76s^2$	fcc	3.84	8.2	$\parallel$
Pt	$5d^96s^1$	fcc	3.92	10.4	$\perp$
Rh	$4d^85s^1$	fcc	3.80	7.0	$\parallel$
W	$5d^46s^2$	bcc	3.17	26.3/3.1	$\parallel$

* the lattice misfit along Co[11$\bar{2}$0]$\parallel$W[001] and along Co[1$\bar{1}$00]$\parallel$W[1$\bar{1}$0] is around 26.3% and 3.1%, respectively [86].

different systems along with the results in this study are compared and summarized here.

For Co/W(110) and Co/Pt(111) systems, the strain relaxation of the Co film occurs at a low Co coverage, due to the substantial lattice mismatch between Co and W(110) and Pt(111) surfaces. Co films on W(110) and Pt(111) remain less strained when two possible hcp- and fcc-like stacking sequences, i.e., Co(111) and Co(0001), are formed [77, 81, 87, 88]. They are close-packed Co layers with two non-equivalent threefold hollow sites. However, the in-plane lattice distance and the adsorption energy of these two non-equivalent threefold hollow sites are very similar [77, 81, 87, 88].

On the other hand, fcc Co(001) films grow pseudomorphically on fcc Cu(001), Ir(001) and Rh(001) substrates. The Co(001) films grown on Ir(001) exhibit a large tensile strain, in contrast to the nearly lattice matched Co(001) films on Cu(001). In addition, for the Co films grown on different kinds of substrates, the electronic hybridization of the 3d Co films with those of 3d, 4d and 5d nonmagnetic metallic substrates is certainly different. The effect of the different atomic structure of the Co film, i.e., Co(001), Co(111) and Co(0001), the epitaxial strain and the interfacial electronic hybridization on the magnetic excitations is discussed in detail as follows.

First of all, we focus on the lowest-energy magnon mode ($n = 0$). Figure 5.1 summarizes the measured magnon dispersion relation of the "acoustic" magnons ($n = 0$) in Co, Fe and FeCo films grown on different substrates. As shown in Fig. 5.1, the similar magnon energies and the dispersion relation is held quite well for the Co films grown on different substrates, meaning that the properties of the "acoustic" magnons do not substantially depend on the substrates (e.g., interfacial electronic

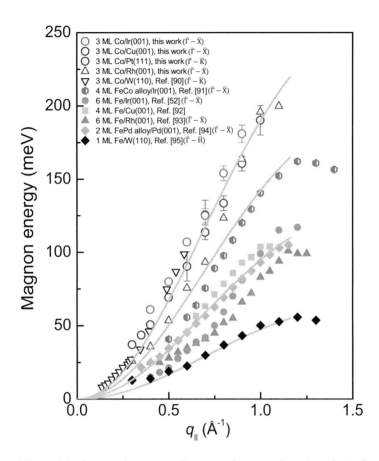

Figure 5.1: A comparison among the measured magnon dispersion relation for the Co, FeCo alloy and Fe films grown on different substrates. The measured magnon dispersion relation along the main symmetry axes of the surface Brillouin zone in each system is indicated in the diagram legend. The symbols represent the experimental data. The solid gray lines indicate the general trend as a guide to the eyes. The relative relation of the gray lines is plotted by the ratio of 4:3:2:1.

hybridization), magnetic anisotropy, epitaxial strain, surface orientation of the film, and substrate surface reconstruction. Such an observation has also been reported earlier on film thicknesses larger than 3 ML [89, 90].

To understand the dominating effects on the nature of the "acoustic" magnons in ferromagnetic transition metals, the magnon dispersion relation obtained on 4 ML FeCo/Ir(001) [91], 6 ML Fe/Ir(001) [52], 4 ML Fe/Cu(001) [92], 6 ML Fe/Rh(001) [93], 2 ML FePd alloy/Pd(001) [94] and 1 ML Fe/W(110) [95] is presented as well in Fig. 5.1.

We find that the magnon energies in Fe films are pronouncedly reduced by half, compared to the ones of the Co films, in spite of the fact that Fe has a larger atomic magnetic moment than the one of Co [22]. The magnon energies of the FeCo film are located between the one of Co and Fe films. Moreover, for a single monolayer Fe film, only half of the magnon energies is observed as compared to that of the Fe film with a thickness larger than 2 ML. We find that the relation of the magnon energies for Co films: FeCo alloy: Fe films (> 2 ML): 1 ML Fe film is about 4:3:2:1, which is represented by the gray lines in Fig. 5.1.

Correspondingly, the experimentally observed spin-wave stiffness constant D of bulk fcc Co is around 580 meVÅ^2, which is around a factor of two larger than the one of bulk bcc Fe (280 meVÅ^2) [96]. Apart from the effect of the coordination number,[1] for Fe bulk and thin films, an effectively reduced magnitude of the spin-wave stiffness constant arises from the antiferromagnetic coupling of the 2nd, 3rd or 4th nearest-neighbors [97]. In contrast, the ferromagnetic coupling in Co bulk and thin films is predominated. The exchange coupling between the next-nearest neighbors (and beyond) is too small and can be neglected. This fact is a consequence of a greater overlap (or larger spatial extent) of the $3d$-wavefunctions of the Fe atoms, as compared to the relatively localized $3d$ wavefunctions on the Co atoms. It is known that the magnon energy of the "acoustic" mode highly depends on the average exchange parameters in the system. Thus, the presence of the antiferromagnetic coupling in Fe may give rise to the average lowered exchange parameters and softened magnon energies. That also could explain the observed magnon energies in the FeCo alloy film being close to the average value of the ones in the pure Co and Fe films.

The dependence of the exchange interaction and the coordination number on the magnon energies in the thin ferromagnetic film will be explicitly interpreted in detail in following sections. A comparison among the results of Co, Fe and FeCo films

---

[1]The number of the (next-) nearest neighbors in the fcc and bcc crystal is 12 (6) and 8 (6), respectively. However, it is important to note that the distance of the next-nearest neighbors of the bcc structure could be smaller than the one of the fcc structure. Thus, depending on the range and nature of the wavefunctions, the effective coordination number in the bcc structure can be larger than the one in fcc structure by including the next-nearest neighbor shell (and beyond) into the bonding.

on different substrates reveals the fact that the "acoustic" magnon mode is strongly element dependent.

5.1.2   Confined magnon modes in the 3 ML Co films

For a 3 ML thick ferromagnetic film grown on a non-magnetic substrate, one would expect that the atomic bonding environment for atoms situated at the surface, middle and interface layers is substantially different from each other, due to the presence of the distinct substrate/film and vacuum/film interfaces. For instance, the lowered coordination number at the film surface leads to an enhanced magnetic moment [22, 23]. At the interface, however, this effect may be counteracted by strong electronic hybridization between film and substrate that may lead to the suppression of the ferromagnetism [24, 25].

Each layer of a "3 ML" film encounters a different environment, so that the physical quantities such as the exchange interaction are layer dependent. The properties of the confined magnon modes depend on the vacuum (surface) and the substrate (interface) boundaries. Each magnon mode therefore contains different contributions from the surface, the interior, and the interface layer. In other words, different layers have a different magnitude of the contribution to each magnon mode. The magnon modes with the quantum number of $n = 0$, 1 and 2 have a preference (i.e., a larger amplitude) to localize at the interface, surface and interior layers, respectively. Thus, they are usually referred to the interface, surface and interior magnon modes, as will be discussed in detail in Sec. 5.2. Since the thickness of a 3 ML film is smaller than the mean free path of electrons in Co, they can all be excited and probed by low-energy electrons.

As shown in Fig. 5.2, in the case of the magnon modes with $n = 0$ and 1, the dispersion relation is very similar for all systems. However, in the case of the highest-energy magnon branch ($n = 2$), a qualitatively different behaviour is observed. While, for the highest-energy mode ($n = 2$), the magnon energies for Co/Ir(001) and Co/Cu(001) are almost identical up to 0.5 Å, a substantially smaller energy is found for Co/Pt(111). For instance, at $\triangle K_\parallel = 0.3$ Å^{-1}, the energies for Co/Pt(111) and Co/Ir(001) are around 290 and 410 meV, respectively. In particular, the energy ratio between Co/Pt(111) and Co/Ir(001) near the center of the Brillouin zone is around three to four.

In the following, the exchange interaction between atoms within the same atomic layer is denoted as *intralayer* exchange interaction ($J_\parallel$) and the one between the neighboring layers is referred to as *interlayer* exchange interaction $J_\perp$. The annotations of the nearest-neighbor (NN) intra-layer ($J_\parallel$) and inter-layer ($J_\perp$) exchange interaction for three-atomic-layer-thick fcc(001) and fcc(111) crystals are shown in

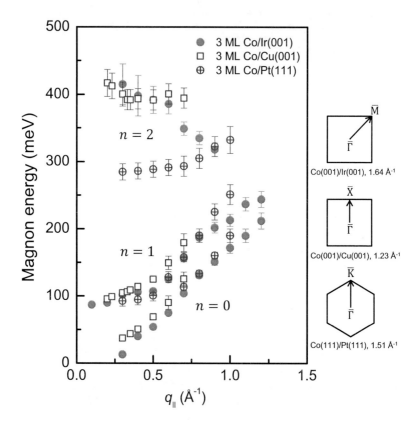

Figure 5.2: The magnon dispersion relation of 3 ML Co films grown on Ir(001) (●), Cu(001) (□) and Pt(111) (⊕) along the $\overline{\Gamma} - \overline{M}$, $\overline{\Gamma} - \overline{X}$ and $\overline{\Gamma} - \overline{K}$ directions of the surface Brillouin zone, respectively. The length of the symmetry axis of the surface Brillouin zone for each system is noted in the right side of the figure.

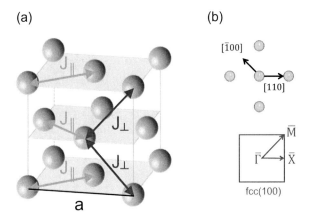

Figure 5.3: (a) Nearest-neighbor intra-layer ($J_\parallel$) and inter-layer ($J_\perp$) exchange interaction for a three-atomic-layer-thick fcc(001) crystal with the lattice constant a. (b) High symmetry directions of the fcc(001) surface in real space (top), and the corresponding surface Brillouin zone in reciprocal space (bottom).

Figs. 5.3 and 5.4, respectively. As discussed in Sec. 2.1.3, the energies of confined magnon modes at $q_\parallel = 0$ are only dominated by the interlayer exchange interaction ($J_\perp$) in thin magnetic films. Assuming a uniform $J_\perp$ in both systems, the ratio of the number of the interlayer nearest-neighbor between the fcc(111) and fcc(001) films is $\frac{3}{4}$ (see Figs. 5.3 and 5.4) and, however, the total nearest-neighbor number is the same (12) in both cases.[2] Remarkably, for the highest-energy mode, the energy ratio of $\frac{3}{4}$ between Co/Pt(111) and Co/Ir(001) near the center of the Brillouin zone scales quite accurately with the interlayer nearest-neighbor coordination.

In contrast to the highest-energy mode, the energy of the second lowest energy magnon mode ($n = 1$) of Co/Pt(111) is very similar to Co/Ir(001) (Fig. 5.2). This mode is much less determined by the interlayer exchange interaction, but the intralayer interaction becomes more important. This can be easily understood from the classical description that the low-energy confined magnon mode has smaller phase deviation of precessing spins along the surface normal. As we will discuss in the next section, the highest energy magnon mode is particularly sensitive to the interlayer exchange interaction.

We would like to emphasize that such a comparison provides just a rough estimation. The energy of the high wave-vector magnons is correlated highly with the number of neighbors and the corresponding strength of the exchange interaction. The

---

[2]The interlayer (intralayer) nearest-neighbor number for 3 ML fcc(001) and fcc(111) crystals is 8 (4) and 6 (6), respectively. The total nearest-neighbor number in both systems is 12.

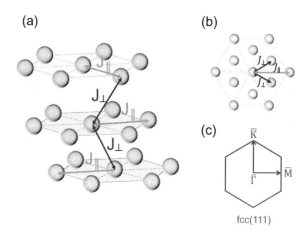

Figure 5.4: (a) Nearest-neighbor intra-layer ($J_\parallel$) and inter-layer ($J_\perp$) exchange interaction for a three-atomic-layer-thick fcc(111) crystal. The ABC stacking sequence of the subsequent (111) planes is presented by green, orange and blue color. (b) Projection of the 3 ML fcc(111) crystal with pathways of the intra- and inter-layer exchange interaction. Two types of the threefold hollow sites are occupied by the layers above and below. (c) High symmetry directions, $\overline{\Gamma} - \overline{K}$ ($[1\bar{1}0]$) and $\overline{\Gamma} - \overline{M}$ ($[11\bar{2}]$), of the fcc(111) surface Brillouin zone.

magnon softening cannot be explained only by a reduction of the number of nearest-neighbors. One has to take the details of the electronic structures, the hybridization effect between films and substrates and the effective layer-dependent interatomic exchange parameters into consideration. These effects will be considered in more detail later in Sec. 5.3 and 5.4.

In Fig. 5.2, the "downward", "flat", and "upward" dispersion relation of the highest-energy magnon branch for the 3 ML Co film on Ir(001), Cu(001) and Pt(111) substrates is experimentally revealed. Correspondingly, the opposite sign of the group velocity $[v_g = \partial_{K_\parallel} E(K_\parallel)]$ of Co/Ir(001) ($v_g < 0$) and Co/Pt(111) ($v_g > 0$) is obtained, while the nearly "flat" dispersion relation of Co/Cu(001) reflects the nearly zero v_g. By comparing the highest-energy magnon mode ($n = 2$) of Co/Pt(111) and Co/Ir(001), one realizes that even though the magnon energy near the center of the Brillouin zone in the case of Co/Pt(111) is smaller, the magnon energy of these two systems reaches the same value at around 0.9 Å^{-1}.

Our results highlight that, apart from the quite robust low-energy modes, the highest-energy magnon mode is substantially sensitive to the choice of the substrate and the atomic structure of the Co films. For the highest-energy magnon branch, near the zone center, a direct observation of the significant shift to lower energies of about

120 meV for 3 ML Co/Pt(111), compared to 3 ML Co/Ir(001) and Co/Cu(001), is experimentally observed. It is important to note that the relative strength between the in-plane and out-of-plane exchange interaction plays a decisive role in determining the magnon dispersion relation. Differing from an uniform exchange interaction in a simple cubic structure, the diverse intra- and inter-layer exchange interaction can be induced by the distorted structure. To understand the microscopic origin of the observed distinct dispersion relation in more details, in the following section, we discuss how the magnon dispersion relation influenced by the modifications of the ratio of the intra- and inter-layer exchange interaction in the system. For instance, these considerations will show that the "downward" dispersion relation in Co/Ir(001) arises from a larger effective interlayer exchange coupling, compared to the intralayer one, as a direct consequence of the compressed tetragonal distortion of the Co film.

5.2 Confined magnon modes in the Heisenberg model

In the following, the Heisenberg model shall be used to derive a simple qualitative picture how the dispersion relation of the confined magnon modes depends on intra- and inter-layer exchange parameters. By means of the Heisenberg model, we provide a detailed characterization of the dispersion relation of the confined magnon modes using the example of a 3ML fcc(001) ferromagnetic film ($N = 3$) with isotropic ($J_\parallel = J_\perp$) and anisotropic ($J_\parallel \neq J_\perp$) exchange parameters. Based on this description, the dispersion relation along the $\overline{\Gamma} - \overline{X}$ ([110]) and the $\overline{\Gamma} - \overline{M}$ ([$\bar{1}$00]) directions of the fcc(100) surface is derived. Note that the anisotropy and dipole effects on the high wave-vector magnons are small compared to J and can be neglected.

Figure 5.3 shows the nearest-neighbor (NN) intra-layer ($J_\parallel$) and inter-layer ($J_\perp$) exchange interaction for a three-atomic-layer-thick fcc(001) crystal with the lattice constant a. The dispersion relation along the [110] ($\overline{\Gamma} - \overline{X}$) direction of the three magnon modes with NN intra- ($J_\parallel$) and inter-layer ($J_\perp$) exchange interaction, derived from Eq. (2.22), is given by

$$\hbar\omega_0(q_\parallel) = 4\left[2SJ_\parallel - 2SJ_\parallel\cos^2\left(q_\parallel\frac{\sqrt{2}}{4}a\right) + 3SJ_\perp - SJ_\perp\sqrt{1 + 8\cos^2\left(q_\parallel\frac{\sqrt{2}}{4}a\right)}\right] \quad (5.1)$$

$$\hbar\omega_1(q_\parallel) = 8\left[SJ_\parallel - SJ_\parallel\cos^2\left(q_\parallel\frac{\sqrt{2}}{4}a\right) + SJ_\perp\right] \quad (5.2)$$

$$\hbar\omega_2(q_\parallel) = 4\left[2SJ_\parallel - 2SJ_\parallel\cos^2\left(q_\parallel\frac{\sqrt{2}}{4}a\right) + 3SJ_\perp + SJ_\perp\sqrt{1 + 8\cos^2\left(q_\parallel\frac{\sqrt{2}}{4}a\right)}\right] \quad (5.3)$$

where $\hbar\omega_0$, $\hbar\omega_1$, and $\hbar\omega_2$ correspond to the energies of the magnon modes with the quantum number of $n = 0, 1$, and 2, respectively. For each particular magnon mode, the magnon amplitude A_N in each atomic layer is

$$n = 0: \ A_\mathrm{I} = A_\mathrm{III} = A, \ A_\mathrm{II} = \frac{\sqrt{1 + 8\cos^2\left(q_\parallel\frac{\sqrt{2}}{4}a\right)} - 1}{2\cos\left(q_\parallel\frac{\sqrt{2}}{4}a\right)}A \quad (5.4)$$

$$n = 1: \ A_\mathrm{I} = -A_\mathrm{III} = A, \ A_\mathrm{II} = 0 \quad (5.5)$$

$$n = 2: \ A_\mathrm{I} = A_\mathrm{III} = A, \ A_\mathrm{II} = -\frac{\sqrt{1 + 8\cos^2\left(q_\parallel\frac{\sqrt{2}}{4}a\right)} + 1}{2\cos\left(q_\parallel\frac{\sqrt{2}}{4}a\right)}A \quad (5.6)$$

A_I and A_III refer to the magnon amplitude of the two surfaces at the distinct sides of the film, and A_II to the one at the central layer.

Particularly, at $q_\parallel = 0$, the energies of confined magnon modes based on Eqs. (5.1)–(5.3) are

$$\hbar\omega_0(0) = 0 \quad (5.7)$$

$$\hbar\omega_1(0) = 8\ SJ_\perp \quad (5.8)$$

$$\hbar\omega_2(0) = 24\ SJ_\perp \quad (5.9)$$

The magnon amplitude of each layers with respect to the magnon energies of $\hbar\omega_0(0)$, $\hbar\omega_1(0)$ and $\hbar\omega_2(0)$ is

$$n = 0: \ A_\mathrm{I} = A_\mathrm{II} = A_\mathrm{III} = A \quad (5.10)$$

$$n = 1: \ A_\mathrm{I} = -A_\mathrm{III} = A, \ A_\mathrm{II} = 0 \quad (5.11)$$

$$n = 2: \ A_\mathrm{I} = A_\mathrm{III} = A, \ A_\mathrm{II} = -2A \quad (5.12)$$

which are obtained from Eqs. (5.4)–(5.6) at $q_\parallel = 0$. For the lowest-energy mode ($n = 0$), as the Goldstone theorem demands [98], in the absence of external magnetic field, a zero-energy at $q_\parallel = 0$ is found, i.e., $B_{ext} = 0$ and $\hbar\omega_0 = 0$ [Eq. (5.7)], in which all spins precess in-phase with the identical amplitude in all three atomic planes [Eq. (5.10)]. With increasing wave vector, we obtain a relatively large amplitude of the lowest energy mode ($n = 0$) at the two surface layers [Eq. (5.4)].

It is important to note that, for a simple-cubic 3 ML film, the atomic planes are stacked directly above the one in the layer beneath. Namely, the nearest-neighbor atoms between neighboring layers simply lie on a straight line along the direction perpendicular to the surface plane. For the lowest-energy mode ($n = 0$), due to the fact that no phase difference between precessing spins along the surface normal direction, the characteristics of the lowest-energy magnon mode are governed by the exchange interaction between atoms within the same atomic layer ($J_\parallel$). Thus, for a simple-cubic 3 ML film with the uniform $J_\parallel$, the magnon amplitude of the lowest-energy mode ($n = 0$) is equivalent in all layers, and does not depend on the wave vector, i.e., $A_I = A_{II} = A$ for all $q_\parallel$.

However, for a close-packed fcc 3 ML film, the atomic planes are shifted relative to one another. The nearest-neighbor inter-layer exchange interaction ($J_\perp$) can be decomposed into in- and out-of-plane components. Namely, $J_\perp$ actually contains a nonzero in-plane component (see Fig. 5.3). The magnitude of the effective in-plane component of $J_\perp$ depends on the interlayer coordination number of each layer. Therefore, in addition to the primarily uniform $J_\perp$, the effective in-plane component coming from $J_\perp$ in the interior layer is larger than the one in two surface layers. It is for this reason that we see the non-uniform magnon amplitude of the lowest-energy mode in the fcc crystals as $q \neq 0$. Above interpretation clearly demonstrated that the properties of the lowest-energy mode are significantly sensitive to the atomic bonding configuration in the ferromagnetic films.

One can find in Eqs. (5.7)−(5.9) that, for a 3ML film, the magnon energies at $q_\parallel = 0$ increase as the quantum number n increases, being 0, 8 $SJ_\perp$ and 24 $SJ_\perp$ for $n = 0$, 1 and 2, respectively, which are only governed by the inter-layer exchange interaction $J_\perp$. In Fig. 5.5, the schematic diagram indicates the phase relationship of the precessing spins between adjacent layers for the three magnon modes with the quantum number of $n = 0, 1$ and 2. The precessing phase of spins changes from one atomic layer to the other along the surface normal direction even at $q_\parallel = 0$ for the confined magnon modes of $n = 1$ and 2. In other words, the magnon modes of $n = 1$ and 2 do not exhibit a uniform spin precession perpendicular to the film, but instead have the number of one-half and one wavelengths spanned between two surfaces (boundaries), respectively. As a result, the confined magnon modes of $n = 1$ and 2 have finite and non-zero energies even at $q_\parallel = 0$, which increases with the quantum number (n) of the magnon modes.

Based on Eqs. (5.4) and (5.10), the lowest-energy mode ($n = 0$) has an identical magnon amplitude across the film at $q_\parallel = 0$, and with increasing wave vector it gradually evolves into the mode preferably localized at two surface layers (owing to the greater magnon amplitude). As shown in Fig. 5.5, for the second lowest-energy

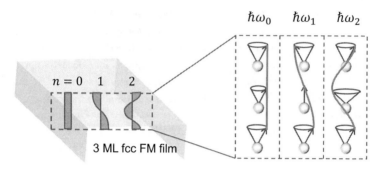

Figure 5.5: Schematic "snapshot" of the magnetization configuration along the surface normal for the magnon modes with the quantum number of $n = 0, 1$ and 2 of a 3 ML fcc ferromagnetic film. Each spin precesses on a cone around the equilibrium magnetization direction ($z-$axis). The corresponding energies $\hbar\omega_n$ of each magnon mode is denoted. For each particular magnon mode, the magnon amplitude in each layer corresponds to the apex angle of a cone. Owing to the relatively large magnon amplitude at certain layers, as $q \neq 0$, the magnon modes of $n = 0$ and 1 are referred to "surface magnon modes," while the highest-energy mode ($n = 2$) is referred to as "central magnon mode."

magnon mode ($\hbar\omega_1$), the spins have the same amplitude with the phase shift of an angle of π in two surface layers. The zero-amplitude at the central layer corresponds to the standing spin wave with one node in the center of the film [Eq. (5.11)]. For the highest-energy mode ($\hbar\omega_2$), spins precess out-of-phase with an angle of π between the central and the surface layers, but precess in-phase with the same amplitude in two surfaces [Eq. (5.12)]. Thus, the highest-energy mode forms two nodes inside of the film.

According to Eqs. (5.4)$-$(5.6), for the lowest- and highest-magnon modes ($\hbar\omega_0$ and $\hbar\omega_2$), a markedly wave-vector-dependent magnon amplitude among different layers is found. For the second mode ($\hbar\omega_1$) the magnon amplitude does not change with the wave-vector. For the two low-energy modes ($\hbar\omega_0$ and $\hbar\omega_1$), the largest amplitude is found at two surface layers. Since the surface localization phenomenon of the two low-energy modes takes place, they are usually called "surface magnon modes." By contrast, for the highest-energy mode ($\hbar\omega_2$), the magnon amplitude in the central layer is larger than the one at the two surface layers. Thus, the highest-energy mode is called "central magnon mode" because it is relatively localized in the center of the film. Since the two surface layers are equivalent, the magnon amplitude at both surfaces in all magnon modes is the same. Namely, the two surface layers contribute equally to all three magnon modes. We point out that the model we present here is particularly simple. In reality, for the film on a substrate, the presence of the

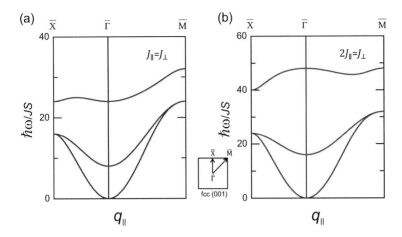

Figure 5.6: The calculated magnon dispersion relation for a three-atomic-layer-thick fcc(001) crystal with (a) isotropic $J_\| = J_\perp$ and (b) anisotropic $2J_\| = J_\perp$ exchange interaction along the $\overline{\Gamma} - \overline{X}$ and the $\overline{\Gamma} - \overline{M}$ directions of the surface Brillouin zone. All diagrams show the first Brillouin zone only.

substrate leads to two "surface" layers which are not equivalent. Then, one of them is the interface layer.

Figure 5.6 shows the calculated magnon dispersion relation based on Eqs. (5.1)−(5.3) along the $\overline{\Gamma} - \overline{X}$ and the $\overline{\Gamma} - \overline{M}$ directions of the surface Brillouin zone. The isotropic case, $J_\| = J_\perp$ (Fig. 5.6a), and the anisotropic case, $2J_\| = J_\perp$ (Fig. 5.6b), are considered and shown. Note that the equations for the magnon dispersion relation along the $\overline{\Gamma} - \overline{M}$ direction are derived in the similar manner as the one along the $\overline{\Gamma} - \overline{X}$ direction (not shown here). We see that the magnon dispersive behavior in the energy-momentum space strongly depends on the relative strength between the intra- and inter-layer exchange interaction. The magnon energies at the zone center ($\overline{\Gamma}$ point) and the high symmetry points of the Brillouin zone ($\overline{X}$ and $\overline{M}$), represented by the effective intra- ($J_\|$) and interlayer ($J_\perp$) exchange interaction, are listed in Tab. 5.2.

For both $J_\| = J_\perp$ and $2J_\| = J_\perp$ cases, the two low-energy magnon branches show the typical "parabolic" dispersion relation. The magnon energies increase with the in-plane wave-vector. The energy splitting between these two low-energy branches decreases with the wave-vector and they reach the same energy at the zone boundary. We now focus on the highest-energy branch. Remarkably, the magnon energy and the dispersive behavior for the highest-energy mode is much more sensitive to the change of the relative strength of the intra- and inter-layer exchange interaction, due to the higher-degree of localization in the "central" layer. One may estimate the general

Table 5.2: The magnon energies at zone center ($\overline{\Gamma}$) and the high symmetry points of the Brillouin zone ($\overline{X}$ and $\overline{M}$) represented by the effective intra- ($J_\parallel$) and inter-layer ($J_\perp$) exchange interaction for a 3 ML fcc(001) crystal, obtained from Eqs. (5.1)–(5.3).

	$\overline{\Gamma}$	$\overline{X}$	$\overline{M}$
1st mode ($n = 0$)	0	$8(SJ_\parallel + SJ_\perp)$	$8(2SJ_\parallel + SJ_\perp)$
2nd mode ($n = 1$)	$8SJ_\perp$	$8(SJ_\parallel + SJ_\perp)$	$8(2SJ_\parallel + SJ_\perp)$
3rd mode ($n = 2$)	$24SJ_\perp$	$8(SJ_\parallel + 2SJ_\perp)$	$8(2SJ_\parallel + 2SJ_\perp)$

shape of the dispersion by the relative strength of $J_\perp$ and $J_\parallel$. As given in Tab. 5.2, at the $\overline{\Gamma}$ point the the magnon energy is governed only by interlayer exchange interaction, being 24 $J_\perp$. At the $\overline{X}$ point at the zone boundary, the magnon energy is given by 8 $SJ_\parallel + 16\ SJ_\perp$, while it turns to be 16 $SJ_\parallel + 16\ SJ_\perp$ at the $\overline{M}$ point.

The resulting dispersion relation reflects the energy difference between the zone-center ($\overline{\Gamma}$) and the zone boundary ($\overline{X}$ or $\overline{M}$). In other words, the resulting "downward" ("upward") dispersion relation corresponds to the higher (smaller) magnon energy at the zone center (the $\overline{\Gamma}$-point) than the one at the zone boundary. Correspondingly, for the case of $J_\parallel = J_\perp$ (Fig. 5.6a), the "flat" and "upward" dispersion along $\overline{\Gamma} - \overline{X}$ and $\overline{\Gamma} - \overline{M}$ directions are obtained. By contrast, the dispersion in these directions becomes "downward" and "flat" for the case of $2J_\parallel = J_\perp$ (Fig. 5.6b). The above results clearly indicate that the magnon energies and their dispersive feature significantly depend on the direction of the magnon propagation with respect to the symmetry directions of the crystal lattice and the relative strength of $J_\perp$ and $J_\parallel$, particularly for the highest-energy mode.

5.3 Layer-dependent exchange parameters

In order to better understand the effect of the substrates and the tetragonal distortion on the magnetic excitations of the studied systems, first-principles adiabatic spin dynamics calculations based on density functional theory (DFT) are performed by Arthur Ernst in the theory department at the Max-Planck-Institute of Microstructure Physics. The idea behind such an approach is to determine the magnetic ground-state at $T = 0$. In this picture, a magnon is an excitation, where the local magnetic moments deviate slightly from the ferromagnetic ground state (or any other magnetic ground state). The description of spin waves and the magnetic properties at finite temperature is then done using an effective Heisenberg model applied to semi-classical spins. Consequently, the effective exchange parameters between any pair of magnetic moments as well as the estimated magnon dispersion relation over the whole Brillouin

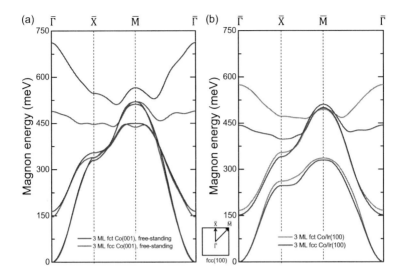

Figure 5.7: The calculated magnon dispersion relation for four different cases:
(i) a freestanding 3 ML face-centered-cubic (fcc) film; (ii) a freestanding 3 ML
face-centered tetragonal (fct) Co film; (iii) a pseudomorphically grown 3 ML fcc
Co film on Ir(001); (iv) a pseudomorphically grown 3 ML fct Co film on Ir(001).
(a) A comparison between a freestanding 3ML fcc and a fct Co film (i and ii). (b)
A comparison between a 3ML fcc and a fct Co film on Ir(001) (iii and iv). The
structural parameters for the four cases are shown in Fig. 5.8.

zone are obtained by an *ab initio* calculation.

In order to understand the role of the substrate and the atomic relaxation of the
Co film, calculations were preformed for four different cases. In the first two sets of
calculations, freestanding face-centered-cubic (fcc) and face-centered tetragonal (fct)
Co films are modeled,[3] while in the other two calculations the same films on Ir(001)
are considered. Due to the pseudomorphic growth of Co on Ir(001), the in-plane
lattice constant for all cases was fixed to the lattice constant of the Ir substrate
being 3.84 Å (Fig. 5.8). For the fct Co films (Fig. 5.8a, c), the interlayer spacing
of all Co layers is 1.61 Å, taken from the experimental data of Refs. [55, 58]. For
all fcc Co films, the out-of-plane lattice constant was chosen to be the same as the
in-plane one (Fig. 5.8b, d). The interlayer distance between Co and Ir is fixed to
be 1.78 Å, according to Ref. [55, 58] (Fig. 5.8c, d). The different structural models
serve as the input for self-consistent calculations of the electronic structure within

---

[3]In the crystallographic unit cell, we define a ratio between the vertical (c) and the in-plane
(a) lattice constant as the c/a ratio. In accordance with the Bain transformation path [99], one can
describe the transformation from fcc to bcc, i.e., the c/a ratio between 1 and $\frac{1}{\sqrt{2}}$, as the face-centered
tetragonal (fct) lattice.

the framework of a generalized gradient approximation of density functional theory [100]. The Korringa–Kohn–Rostoker Green's function method is adopted, particularly designed for layered semi-infinite systems [101]. The interatomic exchange constants are obtained by employing the so-called magnetic force theorem, implemented within the Green function method [102].

Freestanding Co films

Figure 5.7a presents a comparison of the calculated magnon dispersion relation between the freestanding fcc and fct Co films. Surprisingly, the two low-energy branches are very similar. In contrast to the two low-energy modes ($n = 0$ and 1), for the highest-energy branch ($n = 2$), the substantially higher energy and the appearance of the "downward" dispersion relation in the $\overline{\Gamma} - \overline{X}$ and the $\overline{\Gamma} - \overline{M}$ directions are found in the case of the fct Co film. Such a qualitative difference in the dispersion of the highest-energy mode is discussed in Sec. 5.2 as a result of the different intra- ($J_{\parallel}$) and inter-layer ($J_{\perp}$) exchange interaction.

The layer-resolved values for the freestanding fcc and fct Co films are shown in Fig. 5.8a and b, respectively.[4] As the interlayer distance of the Co layers changes from 1.92 Å to 1.61 Å, the ratio of the average inter- ($\overline{J_{\perp}}$) and intra-layer ($\overline{J_{\parallel}}$) exchange parameters increases by a factor of 2. The anisotropic in- and out-of-plane exchange parameters in the fct film arise from the significantly anisotropic charge distribution. Thus, we attribute the notably steep "downward" dispersion relation and the considerably higher energy of the highest-energy mode ($n = 2$) in the fct Co film to the substantially enhanced interlayer exchange interaction induced by the lattice compression (Fig. 5.7).

The enhanced interlayer exchange interaction with the decreasing interlayer distance can be associated with a larger overlap of states between atomic layers. However, the exchange parameters within the atomic layer for the fct Co film are smaller than the one of the fcc Co films, in spite of the fact that the in-plane lattice constant for both fcc and fct films is the same. We suggest that the compressed Co film provokes a substantial spin density redistribution in the system. A very similar phenomenon is also observed for Fe(111) films [103]. The weaker intralayer exchange interaction of the fct film compared to the one of the fcc film may result from the compensation for the much greater overlap of states between layers. However, the average exchange parameters ($\overline{J_{\perp,\parallel}}$) are almost the same in both freestanding films. This explains the almost identical dispersion of the two low-energy modes in both free-standing films. The insensitivity of the two low-energy modes to the relative strength of the intra- and inter-layer exchange parameters, as already discussed based on the Heisenberg

---

[4]We note that in Co, only the exchange interaction between the nearest neighbors is important. The interaction between next-nearest neighbor (and beyond) is too weak and can be neglected.

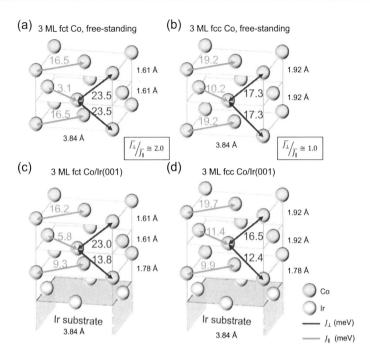

Figure 5.8: The calculated interatomic exchange parameters for four different cases: (a) a freestanding 3 ML face-centered tetragonal (fct) film, (b) a freestanding 3 ML face-centered-cubic (fcc) Co film, (c) a pseudomorphically grown 3 ML fct Co film on Ir(001) and (d) a pseudomorphically grown 3 ML fcc Co film on Ir(001). Only the nearest-neighbor intra- and interlayer exchange parameters are denoted. The values are given in meV.

model in Sec. 5.2, is confirmed.

Here, the change of the effective exchange parameter from $J_\parallel = J_\perp$ to $2J_\parallel = J_\perp$, as illustrated in Sec. 5.2, can be understood as the consequence as the decreasing interlayer spacing. In particular, the case of an isotropic exchange parameter ($J_\parallel = J_\perp$) in the Heisenberg model corresponds to the fcc freestanding film, while the anisotropic one ($2J_\parallel = J_\perp$) corresponds to the fct one. Notably, we see the dispersion features obtained on the basis of first-principles calculations agree qualitatively well with the results that we already derived using the Heisenberg model in Sec. 5.2. Therefore, even without performing the first-principles calculations, one can predict the general trend of the dispersion relation of the confined magnon modes and estimate the relative strength between the effective in- and out-of-plane exchange interaction in the system by means of a simple Heisenberg model. With the help of the first-principles calculations, the effective interatomic exchange interaction between each pair of magnetic moments in the system can be quantified. This further allows to address the effect of the substrates on the magnons in the ultrathin ferromagnets, as will be discussed in the following.

Surface and interface modes

Figure 5.7b shows a comparison between the calculated magnon dispersion relation for the 3 ML fcc and fct Co films on Ir(001). Remarkably, the energies and the dispersion relation of the second lowest energy mode remain almost the same as the ones of the freestanding films, while the lowest- and highest-energy modes are lowered in energy. Especially for the lowest-energy mode, the energies of both fcc and fct films on the substrate are largely reduced by the same scale ($\cong$ 110 meV at $\overline{\mathrm{M}}$), compared to the ones of the freestanding films. For the highest-energy mode, a more pronounced reduction in energy for fct Co/Ir(001) ($\cong$ 140 meV at $\overline{\Gamma}$) than the one for fcc Co/Ir(001) ($\cong$ 50 meV at $\overline{\Gamma}$) is found. These findings clearly demonstrate that the Ir-substrate-induced renormalization of the magnon energies in Co films is significant and is remarkably mode- and wave-vector dependent. The reduced energies of the highest-energy mode for the compressed tetragonal Co film on Ir substrate arise from the stronger interfacial Co_{3d}-Ir_{5d} electronic hybridization.

Figures 5.8c and d present the corresponding exchange parameters of the fct and fcc Co films on Ir(001), respectively. By comparing the values to the ones of the freestanding films, we find that, for both fcc and fct films on the substrate, the exchange parameters in the Co layer adjacent to the Ir substrate are substantially reduced, reflecting the states of the Co films hybridize strongly with the states of the Ir substrate. For both fcc and fct Co films on the Ir substrate, the same amount of energy reduction of the lowest-energy mode is a consequence of the weakened exchange interaction in both systems. However, according to Tab. 5.2, the contribution of the interlayer exchange interaction ($J_\perp$) in the magnon energy is higher for the highest-energy magnon mode than that for the two low-energy modes. We find that the considerably reduced magnon energy of the highest-energy mode for fct Co/Ir(001), compared to its freestanding counterpart, originates from a rather sharp decrease of the interlayer exchange parameters. First-principles calculations reveal that the presence of the Ir substrate leads to a pronounced magnon-softening of the highest-energy magnon mode for the tetragonally distorted film.

We emphasize that, for the freestanding films, we find that the two low-energy modes are degenerate at the Brillouin zone boundary. This can be easily understood, as both surface layers with the same magnon amplitude (see Sec. 5.2) contribute equally to the two low-energy modes. After, including the substrate, the two "surface" layers are not equivalent anymore. The presence of the substrate lifts the degeneracy

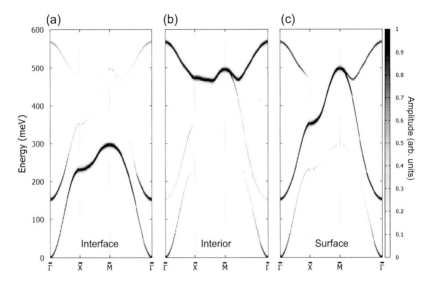

Figure 5.9: The susceptibility Bloch spectral function projected onto the (a) interface, (b) interior and (c) surface layers for 3 ML Co/Ir(001). The scaling of the amplitude of the spectral function is given by the colour bar. The maximum and minimum amplitude is scaled by 0 and 1, respectively.

of the two low-energy modes (see Fig. 5.7). The relatively weak exchange interaction at the interface layer, arising from the strong interfacial Co_{3d}-Ir_{5d} electronic hybridization, has its consequence on a substantial magnon softening of the lowest-energy mode. In other words, the softened lowest-energy mode is mainly localized at the interface.

In order to specify the localization of the distinct magnon modes, the susceptibility Bloch spectral function of the three magnon modes for 3 ML Co/Ir(001) is calculated. Figures 5.9a−c show the spectral function, projected onto the interface, interior, and surface layers. The normalized amplitude of the spectral function is given by the colour bar. Each magnon mode contains different contributions from the interface, the interior, and the surface layer. The magnons at interface provide a dominant contribution to the magnon mode with $n = 0$ and the magnon mode with $n = 1$ have a preference (i.e., a larger amplitude) to localize at the surface, while the interior magnons mainly contribute to the mode with $n = 2$. Thus, the lowest- to highest-energy magnon branches can be characterized by the interface ($n = 0$), surface ($n = 1$), and interior ($n = 2$) magnon modes, as we interpreted in Sec. 5.2. Because the second lowest-energy mode is mainly coming from the surface, it explains the quite robust character of the second lowest-energy mode, as shown in Fig. 5.7.

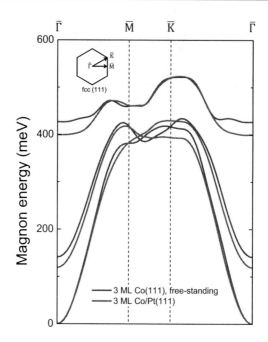

Figure 5.10: A comparison between the calculated magnon dispersion relation of a free-standing 3 ML Co(111) (purple) and such a film on Pt(111) (pink). The in- and out-of-plane nearest-neighbor distance of the Co films in both systems is the same as the one of Pt (2.775 Å).

Interface hybridization and substrate effects

Ir is usually considered as a wide band system due to the spatially more extended $5d$ states. This may result in a strong electronic hybridization between the $3d$ states of the Co film and the unfilled $5d$ states of the Ir substrate. It is known that the electronic hybridization between the ultrathin film and the substrate often leads to considerable modification of the spin-orbit and the exchange coupling in the system that may bring about the unexpected or peculiar magnetic properties of the thin ferromagnetic film [22, 104, 105]. Correspondingly, a large uniaxial magnetic anisotropy and the pronounced magnon softening of the ultrathin Co(001) films grown on the Ir(001) substrate can be attributed to the strong Co_{3d}-Ir_{5d} electronic hybridization at the interface.

This situation is different for Co films grown on the Cu substrate, where the closed $3d$-shell of Cu has rather localized $3d$ states and, thus, shows a relatively narrow d-band lying well below the Fermi level. The states of the Co film hybridize

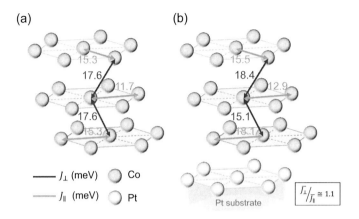

Figure 5.11: The calculated nearest-neighbor interatomic exchange parameters for (a) a freestanding 3 ML Co(111) film and (b) 3 ML Co(111)/Pt(111). The values are given in meV. The in- and out-of-plane nearest-neighbor distance of the Co films is used the same as the one of Pt (2.775 Å). The Co-Pt interlayer distance of 1.8 Å is taken from Ref. [79].

only weakly with the ones of the Cu substrate in the regions around the Fermi level. Correspondingly, the magnon energy shows no significant difference when comparing a Co(001) film on Cu(001) and its freestanding counterpart [15].

However, although both Ir and Pt are $5d$ transition-metal substrates, the substrate-induced renormalization of the magnon energies in Co/Pt behaves differently from Co/Ir. To understand the role of the Pt substrate on the magnon energies and the exchange interaction in ultrathin Co films, the calculated magnon dispersion relation and the nearest-neighbor interatomic exchange parameters of a 3 ML Co film on the Pt(111) substrate, in comparison with that of its freestanding counterpart, are shown in Figs. 5.10 and 5.11, respectively.

For both cases, the in- and out-of-plane nearest-neighbor distance used for the Co films is the same as the one of Pt (2.775 Å). We find the magnon energies and the effective magnetic exchange interaction for Co/Pt and for the freestanding Co film are similar. For more clarification, we also calculated two additional systems by keeping the nearest-neighbor constant the same as the one of bulk hcp Co (2.511 Å) (not shown). Similarly, no notable magnon energy renormalization of the ultrathin Co film induced by the proximity with the Pt substrate is found. In contrast to the strongly layer-dependent exchange interaction in Co/Ir(001), rather "isotropic" exchange parameters in Co/Pt(111) are observed, reflecting the same in- and out-of-plane nearest-neighbor distance of the film. We attribute this result to the different ratio of the intra- and inter-layer nearest-neighbor number in fcc(111) and fcc(001) films, as elucidated in Sec. 5.1.2.

We note that, in spite of the considerable hybridization of the Co $3d$ states with those of the $5d$ states of the Ir and Pt substrates [66, 106],[5] the substrate-induced magnetic anisotropy and magnon energy renormalization respond in an opposite way in the Co/Ir and Co/Pt systems. As shown in Chaps. 4.1.2 and 4.3.2, a large uniaxial in-plane and out-of-plane magnetic anisotropy for Co/Ir and Co/Pt was observed. The effect of the Ir and Pt substrates on the renormalization of the magnon energies of an ultrathin Co film differs significantly.

Numerous studies utilizing both experimental and computational means demonstrated that the Ir and Pt substrates play a dominant role in determining the magnetic anisotropy energy of ultrathin Co films [66, 106–108]. Those results indicate that the significant contribution of the $5d$ substrates to the magnetic anisotropy energy mainly ascribes to the large spin-orbit coupling (SOC) of the substrate (due to the high atomic number Z) and the induced magnetic moments of $5d$ atoms by the ferromagnetic atoms [66, 106–108].[6] The size and the direction of the magnetic anisotropy of the ferromagnetic film are dominated by the energy difference between the in-plane (xy and $x^2 - y^2$) and out-of-plane (z^2, xz and yz) d-orbitals of the magnetic atoms, which can be modified strongly by the d-orbital states of the substrate [109]. We suggest that the large uniaxial in-plane and out-of-plane magnetic anisotropy for Co/Ir and Co/Pt originates from the differences in d-orbital filling induced by the d states of the substrate atoms and the orbital moment anisotropy. However, the magnetic anisotropy energy is about two orders of magnitude smaller than the nearest-neighbor exchange coupling energy.

The pronounced reduction of the exchange interaction and the Co magnetic moment in Co/Ir is driven by the strong interfacial Co_{3d}-Ir_{5d} electronic hybridization. In contrast, the enhanced Co magnetic moment mediated by the Pt substrate (2.03 μ_B vs 1.93 μ_B for the freestanding surface layer) as well as the almost unchanged interatomic exchange interaction in Co/Pt are revealed. The calculated magnetic moment per Co atom in the surface, interior and interface layers for 3 ML Co(001)/Ir(001) is 1.9 μ_B, 1.8 μ_B, and 1.8 μ_B, while the values for 3 ML Co(111)/Pt(111) are 1.9 μ_B, 1.9 μ_B, and 2.0 μ_B, respectively. As a result, we find that the magnon softening in Co/Ir is substantial, in contrast to only minor variations in magnon energies in Co/Pt (in comparison with the freestanding Co film). Similarly, the magnon dispersion relation and the interatomic exchange interaction of the Co films are influenced weakly by the Cu substrate. The interatomic exchange interaction and magnetic anisotropy

---

[5]In addition, the hybridization-induced magnetic moment of the Ir and Pt substrate atoms adjacent to the Co atoms is comparable ($\cong 0.2$ μ_B).

[6]In contrast, Cu has much weaker spin-orbit coupling. Its induced magnetic moment by Co atoms is too small and can be neglected.

energy of the ultrathin Co film depends crucially on the type of substrates.

5.4   Impact of the electronic structure on the magnons

As mentioned in Sec. 5.3, the magnon dispersion relation and the interatomic exchange parameters for a 3 ML Co film on different substrates are calculated based on *ab initio* density functional theory. The calculated magnon dispersion relation and the spin-resolved density of states (DOS) are presented as the green lines in Fig. 5.12a−c and d−f, respectively. By comparing the experimental and theoretical results, we find that the theoretically calculated magnon dispersion relation agrees qualitatively well with the measured one for a 3 ML Co film on different substrates. However, the quantitative difference of the magnon energies between theory and experiment is still significant. The calculated magnon energy, as compared to the measured one, is overestimated by up to 160 meV for the highest-energy ($n = 2$) mode. For 3 ML Co/Cu(001), the predicted magnon energy, by means of first-principles calculations with atomistic spin dynamics (ASD) simulations as reported in Ref. [110], is even higher in energy than the experimentally observed one, differing by 260 meV. An important aspect that has not yet been included in the discussion, and that might explain this quantitative discrepancy is the electronic structure of the Co film itself.

Although one would expect that, for weakly correlated systems,[7] they are described accurately by DFT methods, it has been claimed that these approaches fail to capture the essential physics of the systems with open d- (and f-) shell, such as the late $3d$ transition metals and the $3d$-transition-metal monoxides. For the prototypical case of Ni, at least three notable differences between experiment and the single-particle DFT-based calculations are known: (i) the measured width of the occupied $3d$ bands is narrower, (ii) the measured "average" exchange splitting is half of the calculated one, and (iii) the LSDA-based calculations cannot reproduce the so-called 6-eV satellite observed in photoemission spectra [14, 19, 111].

The ferromagnetic $3d$ transition metals, Fe, Co and Ni, have incompletely filled d bands in the vicinity of the Fermi surface. The prominent correlation effects and the exchange-split electronic states give rise to the itinerant "ground-state" ferromagnetism in these materials. It has been shown that strong many-body correlation effects in the late $3d$ transition metals result in the narrowing of d bands, and a spin-dependent renormalization of quasiparticle energies and lifetimes [17, 18, 32]. These findings indicate that the single-particle DFT-based approaches are not adequately accurate when applied to the late $3d$ transition metals [17]. These materials have

---

[7]For instance, there are the group IV and II-V semiconductors, *sp*-bonded metals (e.g. Na and Al), etc.

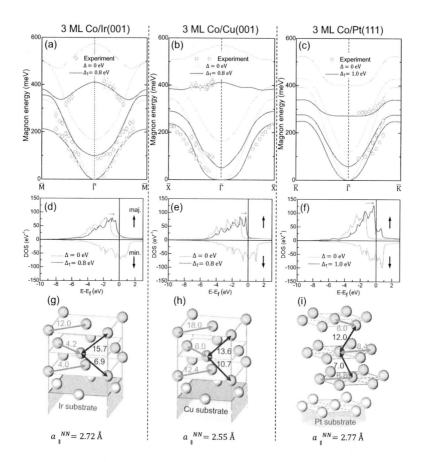

Figure 5.12: (a-c) Calculated (solid lines) and measured (open blue symbols) magnon dispersion relation, (d-f) the spin-resolved density of states and (g-i) the calculated interatomic exchange parameters for Co/Ir(001), Co/Cu(001), and Co/Pt(111) with the film thickness of 3 ML. For a good quantitative agreement with the experimental data, calculations are performed including (i) no spin-dependent potential shift ($\Delta = 0$ eV, green lines of the majority- and the minority-bands), and (ii) a rigid shift of the majority bands towards the Fermi level ($\Delta_\uparrow$, orange lines), while keeping the minority bands unchanged (green lines). For comparison, the calculated magnon dispersion and DOS of (i) and (ii) are plotted together in (a-f). By comparing the measured magnon dispersion relation with the theoretically calculated one, the effective interatomic exchange parameters are quantitatively determined.

lately been denoted as "strong" correlated itinerant ferromagnets [17].[8]

It has been studied theoretically that, for bulk ferromagnets, the spin-wave stiffness constant (D) is extremely sensitive to the "ground-state" exchange splitting of the d bands [112, 113]. For bulk Ni, it was found that D increases rapidly with increasing the size of the exchange splitting (δ), as long as a fully occupied majority band remains in the ferromagnetic ground state (i.e., $\delta \geq 0.35$ eV) [112]. Moreover, it is well known that for bulk Ni the acoustic magnon branch is described well within the LSDA($+U$), whereas the energy of the "optical" mode is significantly too high. The discrepancy of the magnon energy between the experiment and theory has been attributed to an overestimation of the "average" exchange splitting in LSDA($+U$) by a factor of two, compared to photoemission results [14, 114]. By decreasing the "average" exchange splitting to the experimental value, a good agreement of the calculated magnon energies has been obtained [14, 114]. Also for Co films it has be demonstrated that theoretically calculated magnon energies are particularly sensitive to the detailed electronic structure [115].

As described above, the electronic structures of Co cannot be described adequately within the single-particle DFT-based scheme. However, a remarkable agreement with photoemission experiments is found by a relative shift of the majority spin-states of $3d$ bands towards the Fermi level, compared to the almost unchanged minority spin states. This result has been attributed to the stronger correlation effects of the majority spin electrons [18, 32]. The finding leads to the consequence that the separation between the majority and minority-spin states in Co is reduced, as compared to the one predicted by the single-particle DFT-based approaches.

To better understand the influence of the electronic structures on the magnon energies of the ultrathin Co film, we evaluate the effect of the "average" exchange splitting on the confined magnon modes of a 3 ML Co film on different substrates. According to previous photoemission results, the calculations are performed as shown in Fig. 5.12d-f: (i) no spin-dependent potential shift (green lines of the majority- and the minority-bands) for comparison and (ii) a rigid shift of the majority band towards the Fermi level (orange lines), while keeping unchanged the minority band (green lines). The case of the calculation (i) is the same as the one shown in Sec. 5.3. The results of calculated magnon dispersion and DOS are compared in Fig. 5.12a-c and d-f, respectively. As only the majority bands are shifted in calculation (ii), the magnitude of this shift directly corresponds to the reduction of the exchange splitting. Consequently, this leads to a down-scaling of the previously overestimated magnon energies.

---

[8]Indeed, for systems with weak (or "moderate") correlation effects, the physical properties can be described accurately by the single-particle DFT-based scheme which includes the correlation effects through the effective static mean-field interaction. The term of "strong" correlation is generally referred to that the mean-field approximations fail to describe the essential physics in the system [17].

To find the best agreement between experiment and theory, the calculations are performed by a gradual shift of the majority bands in steps of 0.1 eV for each system. We find that the calculated atomic magnetic moment of Co reduces with the decreasing exchange splitting. As shown in Fig. 5.12, for a best quantitative agreement with the experimental data, one would need to shift the majority bands (up-arrow) by 0.8 eV towards to the Fermi level (the orange line) for Co/Ir and Co/Cu, while the minority bands (down-arrow) are not shifted (the green line). Note that one needs a slightly larger shift of 1.0 eV for Co/Pt in order to explain the experimental data very well. Remarkably, the splitting between the highest majority- and minority-spin states of Co d-bands for Co/Pt is 1.06 eV, in good agreement with the photoemission observation of 1.05 eV [116].

It is important to note that it has been previously confirmed by photoemission experiments [18,32] that, the energy renormalization of the majority spin states in Co is $0.7-1.0$ eV. Surprisingly, this finding agrees well with our results. The spin-dependent electronic energy renormalization has been attributed to strong spin-dependent correlation effects [18,19,32]. However, it is expected that the strength of correlation effects in a 3 ML film may be layer dependent. It might also depend on the film structure. This may give rise to a slightly different magnitude of the majority-band-shift for Co/Pt compared to Co/Ir and Co/Cu. To account for the remaining minor quantitative discrepancies between theory and experiment, wave-vector- and layer-dependent correlation effects or more sophisticated many-body calculations are required.

Our results clearly demonstrate that the dependence of the magnon energies on the size of the "average" exchange splitting is dramatic, particularly for the higher-energy magnons (Fig. 5.12a-c). The degree of the renormalized magnon energy for each system is strongly mode- and wave-vector dependent. Correspondingly, the notable modification of the magnon (phase and group) velocities is found that depends on the wave-vector, particularly for the two low-energy magnon modes. However, at the $\overline{\Gamma}$ point, the zero-energy of the lowest-energy mode ($n = 0$) is preserved, in accord with the Goldstone theorem. By comparing the measured and the theoretically calculated magnon dispersion relations, the effective interatomic exchange interaction is quantitatively determined in all systems (Fig. 5.12g-i).

One can clearly see that the weakened interatomic exchange interaction is layer- and system-dependent, although they have a similar energy shift of the majority-spin states. The effective nearest-neighbor exchange parameter is reduced by about $30 - 45\%$, as compared to the unshifted case. This observation indicates that the considerably weakened interatomic exchange interaction is a result of the strong spin-dependent correlation effects in itinerant ferromagnets.

Our results provide convincing evidence that the electronic structure indeed plays a crucial role in determining the magnon energies and the interatomic exchange interaction in ultrathin itinerant ferromagnetic films. We systematically evaluate the effect of the exchange splitting on the energies of the confined magnon modes and the layer-dependent exchange interaction for a 3 ML Co film on different substrates. By a shift of the majority bands of $0.8 - 1.0$ eV towards the Fermi level, the best agreement between the measured and calculated magnon dispersion relations for all systems is obtained. This finding agrees well with previous photoemission results. The strong spin-dependent quasiparticle energy renormalization has been ascribed to the markedly spin-dependent correlation effects in late $3d$ transition metals [17,18,32]. The experimentally observed magnon dispersion relation of the confined magnon modes provides access to the information of the "average" exchange splitting in ultrathin Co films. Our results indicate that many-body correlation effects are essential to describe the magnetic properties and the magnetic interactions in the itinerant ferromagnetic system.

5.5 The magnon lifetime

5.5.1 "Acoustic" magnons in Co and Fe films

Above, the influence of the film structure, the type of substrate and the electronic structure on the energy and the group and phase velocity of the confined magnon modes was discussed for ultrathin Co films on different substrates. Understanding the attenuation mechanism of magnons in the ultrathin films on substrates is an essential issue. In the last section, we discuss the damping and the corresponding relaxation time of magnons in ultrathin Fe and Co films.

To obtain the linewidth and the magnon lifetime, the SPEELS difference ($I_{Diff} = I_- - I_+$) spectra are fitted using a Voigt lineshape, in which the Gaussian part represents the instrumental broadening and the Lorentzian part represents the intrinsic lifetime broadening. The full-width at half-maximum (FWHM) of the magnon peak is interpreted as the inverse lifetime of excitations. The Fourier transform of the Lorentzian in energy (or frequency) domain is an exponential decay in the time domain, $\exp(-t\Gamma/2\hbar)$, where Γ represents the intrinsic linewidth of the Lorentzian peak in energy and $\hbar$ is the reduced Planck constant. The lifetime of the excitation can be derived as $\tau = 2\hbar/\Gamma$ and is usually defined as a time in which the amplitude drops to its e^{-1} value [117].

The linewidth and the lifetime of the acoustic magnon mode ($n = 0$) have been investigated in the following systems: 1 and 2 ML Fe/W(110) [95,117,118], and 2 ML FePd alloy/Pd(001) [94]. To understand the damping effects on the acoustic magnons in Co and Fe films, we compare the intrinsic magnon linewidth and the corresponding magnon lifetime as a function of the magnon energy for 2 and 3 ML Co/Ir(001) with

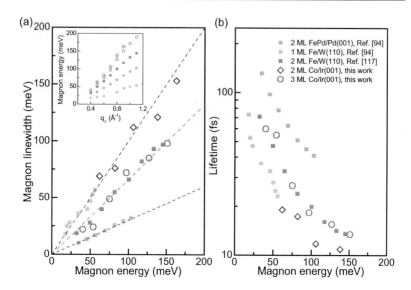

Figure 5.13: (a) The intrinsic magnon linewidth and (b) the magnon lifetime as a function of the magnon energy for ultrathin Co and Fe films grown on different substrates. Blue, orange and green dashed lines are shown as a guide to the eye. The inset in (a) shows the magnon dispersion relation of the acoustic mode ($n = 0$) for all systems. The experimental data displayed in (a) and (b) for each sample system is indicated in the legend of (b).

those previous results, as shown in Fig. 5.13. Note that the magnetic excitations for the 2 ML Co film grown on Ir(001) were also studied in the course of this thesis.

Since the acoustic magnon mode ($n = 0$) involves the uniform precession of spins in the whole system, the presence of a "quantized wave-vector" perpendicular to the film surface has negligible contribution to the damping of acoustic magnons. We clearly see that the intrinsic linewidth and magnon lifetime depends strongly on its energies.

In the adiabatic calculations based on the Heisenberg-Hamiltonian, magnons with an infinite lifetime are characterized by a delta function. However, one of the most distinctive feature of magnons in the itinerant ferromagnets is their proximity to the particle-hole Stoner excitations. This excitations involve the available electronic transition between the occupied majority spin states with crystal wave-vector **k** and the empty minority-spin states with wave-vector **k**+**q**. The wave-vector **k** is associated with the whole Brillouin zone in the system. When the magnon dispersion curve in the momentum-energy space encounters noncoherent particle-hole Stoner excitations, the finite spectral magnon linewidth and lifetime is presented. This mechanism of

the magnon attenuation is usually referred to as the Landau damping. The size of
the linewidth and the corresponding lifetime are associated with the density of the
Stoner states in the corresponding energy and wave-vector range of magnons. At the
corresponding wave-vector q and energy ε of magnons, the possibilities of the available
Stoner transitions which contribute to the damping of magnons have been described
and visualized by the so-called Landau map [13, 15].

The impact of Stoner excitations on the magnon lifetime can vary greatly, depending on the energy and wave-vector of magnons. A higher possibility of "spin-flip"
transitions in the corresponding energy range results in a stronger Landau damping of
magnons. The energy-dependent linewidth and lifetime as shown in Fig. 5.13 reflects
the increase of the Landau damping with increasing excitation energies.

Figure 5.13 shows the strong dependence of the intrinsic magnon linewidth and
its lifetime on the type of the substrates and the film thickness. At certain magnon
energy, the longest magnon lifetime is found in 2 ML FePd/Pd(001), whereas the
shortest one is observed in 2 ML Co/Ir(001) and 1 ML Fe/W(110). It has been
demonstrated that, for an ultrathin magnetic film, the damping of magnons depends
strongly on the details of the electronic hybridization of the states of a ferromagnetic
film and the states of a substrate [13]. The hybridization with the substrate states
may increase (or decrease) the number of the available Stoner transitions near the
Fermi level and thereby enhance (or suppress) the attenuation of magnons.

In particular, for 2 ML FePd/Pd(001), the origin of the weakly damped magnons is
attributed to the relative small density of d-electron states near the Fermi level. Similar to the half-metallic systems, the interfacial electronic hybridization in FePd/Pd(001)
effectively suppresses the possibility of Stoner transitions, resulting in long-living
magnons [15, 94]. The situation is strikingly different in Co/Ir(001) and Fe/W(110).
The electronic hybridization with the states of the Ir and W substrate at the interface
brings about the increase of the available Stoner transitions, thereby results in a larger
decay rate and a sizable reduction of the magnon lifetime.

Moreover, we find that the magnon lifetime decreases with decreasing film thickness. A comparison between 2 and 3 ML Co/Ir(001) clearly demonstrates that, in
spite of a similar magnon dispersion of 2 and 3 ML Co/Ir(001) (see the inset of Fig.
5.13a), the magnon damping in 2 ML Co film is stronger. For 2 ML/Ir(001), a larger
density of d states near the Fermi level is suggested. This is connected with additionally available Stoner transitions near the Fermi level decreasing magnon lifetime in
the 2 ML film.

Interestingly, by comparing the decay rate of 1 ML Fe/W(110) to the one of 2 ML
Fe/W(110), in spite of the fact that the magnon energies are smaller by a factor of
two, the damping for 1 ML Fe/W(110) is more pronounced. Note that, for bulk bcc
Fe, a shorter magnon lifetime has been found, compared to the 1 ML Fe films [15].

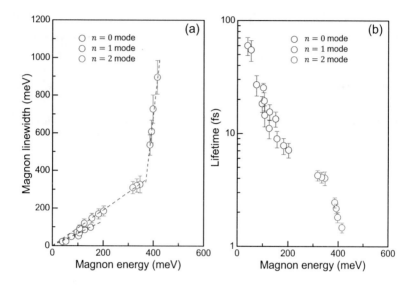

Figure 5.14: (a) The intrinsic magnon linewidth and (b) the magnon lifetime as a function of the magnon energy for a 3 ML Co film on Ir(001). The experimental data of the confined magnon modes corresponding to the quantum number of $n = 0$ (blue), $n = 1$ (green) and $n = 2$ (pink) are presented.

This result reveals the non-monotonous dependence of the magnon attenuation on the dimensionality of the system. It is important to keep in mind that, although the two systems have the same magnon lifetime at the same energy, they possess different wave-vectors and propagation speeds (see the inset of Fig. 5.13a).

5.5.2   Confined magnon modes

Figure 5.14 shows the experimentally observed intrinsic magnon linewidth and lifetime of all three magnon modes for a 3 ML Co film on Ir(001). As mentioned in Sec. 4.1.3, at a certain wave-vector, the intrinsic magnon linewidth becomes broadened with increasing quantum number of the magnon mode.

Based on the degree of the magnon attenuation, three different kinds of magnon decays are distinguished (see Fig. 5.14): (i) For the $n = 0$ mode, the linewidth increases more or less linearly with increasing magnon energies, as presented in Fig. 5.13. (ii) For the $n = 1$ mode, the slope of the linear dependence of the linewidth is slightly increased by around 40%. (iii) For the magnon energy above 370 meV, a simple linear interpolation dose not hold. The magnon linewidth rises steeply as the energy increases. The slope of the linear dependence of the linewidth increases dramatically by a factor of 18, compared to the one of $n = 0$ mode. Correspondingly, the magnon lifetime decreases abruptly above this threshold energy of about 370 meV. The distinct decay rate of different magnon modes is observed. At a certain wave-vector, the weakly- and strongly-damped magnons coexist.

For the low-energy regime, magnons have a well-defined energy and a long lifetime, resulting from a relatively weak Landau damping. The intrinsic linewidth depends monotonously on the magnon energies. In contrast to the low-energy regime, the suddenly broadened linewidth above a magnon energy of around 370 meV suggests that they encounter a high density of Stoner-excitation states. We note that, in different systems, the relative density of the Stoner-excitation states in the corresponding energy and wave-vector regime may differ significantly, leading to a distinct decay rate in each system.

In summary, the linewidth and lifetime of magnons depend strongly on their energies for all magnon modes in an ultrathin ferromagnetic film. A competition between distinct nature of the collective spin-wave and particle-hole excitations plays a decisive role in the magnon Landau damping. It is observed that for a 3 ML Co film on Ir(001), the damping rate of different magnon modes are significantly different.

Chapter 6

Conclusions and outlook

In this work, confined magnon modes and their in-plane dispersion relations in ultra-thin Co films grown on Ir(001), Cu(001), and Pt(111) are investigated by means of spin-polarized electron energy loss spectroscopy. The effect of magnetic anisotropy, epitaxial strain, atomic structures, film thickness, substrate reconstruction, and interfacial hybridization on these magnon modes were investigated.

The magnon mode with the lowest order ($n = 0$) corresponds to the uniform (in-phase) precession of spins in all atomic layers of the film and is called the acoustic mode. In the higher order modes ($n > 0$), spins between adjacent layers are no longer aligned parallel to each other. They are characterized explicitly by the number of n half-wavelength envelopes (and the number of n nodes) of standing spin waves with the total phase shift of $n\pi$ among spins along the out-of-plane direction. Therefore, the number of magnon modes is equal to the number of atomic layers in a film.

All three magnon modes of a 3 ML Co film are probed. The dispersion relation of the highest-order magnon mode, which is characterized by "anti-phase" precessing spins between neighboring layers along the direction perpendicular to the film surface, was measured over a large fraction of the surface Brillouin zone. In contrast to the typical "parabolic" dispersion relations of the two low-energy modes ($n = 0$ and 1), the dispersion relation of the highest-energy mode is entirely different. In particular, a "downward", "flat", or "upward" shape of the dispersion of 3ML Co films on Ir(001), Cu(001) and Pt(111) has been experimentally observed. It was shown that the energy of the $n = 2$ magnon mode near the center of the surface Brillouin zone depends strongly on the interlayer coordination number and the strength of interlayer exchange interaction ($J_\perp$). For instance, the observed energy ratio of $\frac{3}{4}$ between Co/Pt(111) and Co/Ir(001) scales directly with the ratio of the respective interlayer nearest-neighbor coordination numbers. This could be easily understood based on a the classical description.

The magnon dispersion relation of all quantized modes were qualitatively described by using a Heisenberg model. This model provides a way of understanding the interplay between the intra- and inter-layer exchange interaction and their influence on different magnon modes. For instance, for isotropic and anisotropic exchange interaction, the two low-energy magnon branches show the typical "parabolic" dispersion relation for all cases. In contrast, the highest-energy mode ($n = 2$) is much more sensitive to the relative strength of the intra- and inter-layer exchange interaction. For the case of $J_\parallel = J_\perp$, flat and upward dispersion relations of the highest-energy mode are observed, in agreement with the experimental results of Co/Cu(001) and Co/Pt(111), respectively. In these systems, the isotropic exchange parameters are a result of the same in- and out-of-plane nearest-neighbor distance of the film. By contrast, for the case of $2J_\parallel = J_\perp$, a downward dispersion relation is found, which qualitatively agrees well with experimental observations in Co/Ir(001). The anisotropic exchange interaction is a direct consequence of the tetragonally distorted lattice of the Co film grown on Ir(001). Since the highest-energy mode ($n = 2$) is strongly influenced by the layer-dependent variation of the exchange parameter, the changes in layer-dependent electrical and magnetic properties due to interfacial electronic hybridization, epitaxial strain, etc. are reflected directly in the behaviour of this magnon mode.

Analysis of the calculated Bloch's spectral function revealed that different magnon modes can be assigned to different atomic layers. The magnons at the interface (surface) provide a dominant contribution to the (second) lowest-energy magnon mode, while the interior magnons mainly contribute to the highest-energy mode. Accordingly, the distinct magnon branches are characterized as interface ($n = 0$), surface ($n = 1$), and interior ($n = 2$) magnon modes. The confined magnon modes with different quantum number n are associated with different physical properties of ultrathin films, due to different layers where the respective modes are localized. This made it possible to disentangle different effects on the magnetic properties of the ultrathin cobalt films, such as epitaxial strain, film structures, magnetic anisotropy, interfacial electronic hybridization, spin-dependent many-body interactions, and density of Stoner-excitation states.

It was shown that many-body interactions in the electronic states of the ferromagnet lead to a pronounced renormalization of the magnon energies for higher-energy magnons, differing by up to 160 meV. Notably, about 30% reduction in the energies of confined magnon modes is observed. By considering a renormalization of electronic states in the ferromagnet as a shift of the majority bands of 0.8 eV towards the Fermi level, the best agreement between the measured and calculated magnon dispersion relation is found for both Co/Ir and Co/Cu. A slightly larger shift of 1.0 eV was required for Co/Pt. It is clearly shown that the dependence of the magnon energies on the exchange splitting is dramatic, particularly for the higher-energy magnons. The strong spin-dependent quasiparticle energy renormalization has been ascribed to the spin-dependent correlation effects in the Co film, which quantitatively agrees with previous photoemission results [17, 18, 32].

The substrate influences both the magnon energies and its Landau damping. For instance, first-principles calculations for Co films grown on Ir(001) reveal that the presence of the Ir substrate leads to the following results: (i) the lifting of the degeneracy between two low-energy modes at the high-symmetry points and (ii) a weakened exchange interaction between the moments in the layer adjacent to the substrate. Moreover, the hybridization with the substrate states opens up possibilities to enhance or suppress the magnon decay.

The highly spin-, energy-, and momentum-resolved detection of the confined magnon modes in ultrathin ferromagnets opens up new perspectives to the scope of future fundamental and applied studies. From an application point of view, the confined magnon modes at a certain atomic layer can be effectively excited. By tailoring atomic structures and choosing different substrates and capping layers, one would be able to engineer multi-magnon modes with specific frequencies, wavelengths and amplitudes guided inside the planar magnetic film. The exchange-dominated magnon modes describe the dynamics of spins on length scales of ångströms to nanometers and with characteristic timescales of terahertz. By manipulating the amplitude and phase of multi-magnon modes, the proposed non-volatile complex multi-frequency magnonic logic circuits with ultra-low power consumption might be realized [119–123]. Taking advantages of the fact that the distinct magnon modes are spatially localized at different atomic layers might lead to a promising further step towards the realization of reconfigurable atomic-sized magnonic waveguides [124, 125], nano-sized magnonic crystals, and ultimately fast magnetic switching.

From a fundamental point of view, further SPEELS studies of exploring spatially confined exchange-driven magnon modes with higher orders in other ultrathin films are highly desirable, such as strongly correlated compounds, ferroelectrics, Heusler alloys, high-T_C superconductors and frustrated magnets. Such studies will provide a way for experimental investigations of the many-body correlation effects on the magnetic excitations, the layer-dependent exchange interaction, and damping mechanisms in those films to compare the results with the elemental ferromagnet Co. Of particular importance is that the confined magnon modes can serve as a layer-specific probe with both high spatial (ångströms to nanometers) and high temporal (up to attoseconds) resolution. This provides appealing prospects to explore the atomic-layer-resolved properties of exotic states and excitations and to further reveal the interfacial couplings in low-dimensional magnetic systems [126–129].

Appendix A

Description of spin-polarized electrons

The best way to characterize the electron spin is as a form of angular momentum. The spin operators for $S = \frac{1}{2}$ of its projection on the x, y and z axes (z axis here is the quantization axis) in the non-relativistic limit are described by

$$\hat{S}_\alpha = \frac{\hbar}{2}\hat{\sigma}_\alpha \ (\alpha = x, y, z), \tag{A.1}$$

where $\hat{\sigma}_\alpha$ are the Pauli spin matrices, defined as

$$\hat{\sigma}_x = \begin{pmatrix} 0 & 1 \\ 1 & 0 \end{pmatrix}, \hat{\sigma}_y = \begin{pmatrix} 0 & -i \\ i & 0 \end{pmatrix}, \hat{\sigma}_z = \begin{pmatrix} 1 & 0 \\ 0 & -1 \end{pmatrix}, \text{ and } \hat{\sigma}_\alpha{}^2 = \begin{pmatrix} 1 & 0 \\ 0 & 1 \end{pmatrix} = 1 \tag{A.2}$$

Elements of a spin representation are described by a two-component spinor, i.e.

$$\chi = \begin{pmatrix} a \\ b \end{pmatrix} \tag{A.3}$$

where a and b are complex numbers and the normalized state is $|a|^2 + |b|^2 = 1$. The possible observed quantities of the angular momentum along any axis, e.g., $\hat{S}_x$, $\hat{S}_y$ and $\hat{S}_z$, are $+\frac{\hbar}{2}$ and $-\frac{\hbar}{2}$ (e.g., up/down). Here the spinor corresponding to the eigenvalue of $+\frac{\hbar}{2}$ is denoted as χ_+ and the spinor corresponding to the eigenvalue of $-\frac{\hbar}{2}$ is denoted as χ_-. These two spinors are orthogonal to each other, such that $\chi_+^*\chi_- = 0$.

By deriving the equation, $\hat{S}_\alpha\chi_\pm^\alpha = \lambda_\pm\chi_\pm^\alpha$, $(\alpha = x, y, z)$, the eigenstates $\chi_\pm^\alpha$ of the operator $\hat{S}_\alpha$ with the eigenvalues (λ) of $\pm\frac{\hbar}{2}$ are

$$\chi_+^x = \begin{pmatrix} \frac{1}{\sqrt{2}} \\ \frac{1}{\sqrt{2}} \end{pmatrix}, \ \chi_-^x = \begin{pmatrix} \frac{1}{\sqrt{2}} \\ -\frac{1}{\sqrt{2}} \end{pmatrix} \tag{A.4}$$

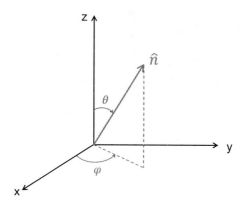

Figure A.1: The representation of the unit vector $\hat{n}$ of the spin direction determined by the polar (θ) and azimuthal (φ) angles in spherical coordinates.

$$\chi_+^y = \begin{pmatrix} \frac{1}{\sqrt{2}} \\ \frac{i}{\sqrt{2}} \end{pmatrix}, \ \chi_-^y = \begin{pmatrix} \frac{1}{\sqrt{2}} \\ -\frac{i}{\sqrt{2}} \end{pmatrix} \tag{A.5}$$

$$\chi_+^z = \begin{pmatrix} 1 \\ 0 \end{pmatrix}, \ \chi_-^z = \begin{pmatrix} 0 \\ 1 \end{pmatrix} \tag{A.6}$$

where $\chi_\pm^x$, $\chi_\pm^y$ and $\chi_\pm^z$ indicate the spin pointing parallel or antiparallel to the x-, y-, and z-axes, respectively. One can examine the conditions $\hat{S}_\alpha \chi_+^\alpha = \frac{\hbar}{2}\chi_+^\alpha$ and $\hat{S}_\alpha \chi_-^\alpha = -\frac{\hbar}{2}\chi_-^\alpha$ are satisfied. Since only the operator $\hat{S}_z$ is diagonal, the representation of the electron spin pointing along the z direction is rather simple.

Now we assume the spin of an electron is along an arbitrary direction, i.e., the unit vector $\hat{n} = \sin\theta\cos\varphi\hat{i} + \sin\theta\sin\varphi\hat{j} + \cos\theta\hat{k}$, determined by the polar (θ) and azimuthal (φ) angles in spherical coordinates. The spin operator $\hat{S}\cdot\hat{n}$ for the component of spins along unit vector $\hat{n}$ with the eigenvalue of $\pm\frac{\hbar}{2}$ can be expressed by

$$\hat{S}\cdot\hat{n} = \sin\theta\cos\varphi\hat{S}_x + \sin\theta\sin\varphi\hat{S}_y + \cos\theta\hat{S}_z \tag{A.7}$$

$$= \frac{\hbar}{2} \begin{bmatrix} \cos\theta & \sin\theta e^{-i\varphi} \\ \sin\theta e^{i\varphi} & -\cos\theta \end{bmatrix} \tag{A.8}$$

Consequently, the eigenstates of χ_+ and χ_- corresponding to the respective eigenvalues of $+\frac{\hbar}{2}$ and $-\frac{\hbar}{2}$ are given by

$$\chi^n_+ = \begin{bmatrix} \cos\left(\frac{\theta}{2}\right) \\ \sin\left(\frac{\theta}{2}\right) e^{i\varphi} \end{bmatrix} \tag{A.9}$$

$$\chi^n_- = \begin{bmatrix} \sin\left(\frac{\theta}{2}\right) \\ -\cos\left(\frac{\theta}{2}\right) e^{i\varphi} \end{bmatrix} \tag{A.10}$$

Both states satisfy the condition of $\chi_\pm \hat{S}\chi_\pm = \pm \left(\frac{\hbar}{2}\right)\hat{n}$ and have the opposite orientation along the direction of the unit vector $\hat{n}$. As a result, the generic spinor χ of an electron with the spin along any direction can be described by a linear combination of two basis vectors (χ_+ and χ_-) in the two-dimensional spin space:

$$\chi = c_+\chi_+ + c_-\chi_- \tag{A.11}$$

where c_+ and c_- are complex numbers. The normalized state is

$$\langle\chi|\chi\rangle = |c_+|^2 + |c_-|^2 = 1 \tag{A.12}$$

For example, $\chi = c_+\chi^z_+ + c_-\chi^z_- = c_+\begin{pmatrix}1\\0\end{pmatrix} + c_-\begin{pmatrix}0\\1\end{pmatrix} = \begin{pmatrix}c_+\\c_-\end{pmatrix}$ can be represented by any spin state. Based on the statistical interpretation of quantum mechanical wavefunctions, the values of $|c_+|^2$ and $|c_-|^2$ are the possibilities to find the states in measurements with spins pointing up and down along the z-axis with the eigenvalue of $+\frac{\hbar}{2}$ and $-\frac{\hbar}{2}$, respectively.

Moreover, for spins oriented in the x direction, the state can be regarded as a superposition of spin states with the opposite z direction carrying the identical amplitude, according to

$$\chi^x_+ = \frac{1}{\sqrt{2}}\begin{pmatrix}1\\1\end{pmatrix} = \frac{1}{\sqrt{2}}\begin{pmatrix}1\\0\end{pmatrix} + \frac{1}{\sqrt{2}}\begin{pmatrix}0\\1\end{pmatrix} = \frac{1}{\sqrt{2}}\chi^z_+ + \frac{1}{\sqrt{2}}\chi^z_- \tag{A.13}$$

Namely, the possibilities one would get $+\frac{\hbar}{2}$ or $-\frac{\hbar}{2}$ for an observation of spin components along the z direction is $\frac{1}{2}$, based on $|c_+|^2 = |c_-|^2 = \frac{1}{2}$. Vice versa, the same possibility is achieved for the spins oriented along z direction but one observes in the x-axis,

$$\chi^z_+ = \begin{pmatrix}1\\0\end{pmatrix} = \frac{1}{\sqrt{2}}\begin{pmatrix}\frac{1}{\sqrt{2}}\\\frac{1}{\sqrt{2}}\end{pmatrix} + \frac{1}{\sqrt{2}}\begin{pmatrix}\frac{1}{\sqrt{2}}\\-\frac{1}{\sqrt{2}}\end{pmatrix} = \frac{1}{\sqrt{2}}\chi^x_+ + \frac{1}{\sqrt{2}}\chi^x_-. \tag{A.14}$$

An electron beam is assembled by individual electrons. A quantum description of an electron beam is based on individual electrons. In practice, by introducing a spin polarization vector ($\mathbf{P}$) to an electron beam, the relative orientation between the polarization of the electron beam ($\mathbf{P}$) and the magnetization direction ($\mathbf{M}$) of the sample is usually signified in experiments. We note that the polarization vector ($\mathbf{P}$) is only denoted as the expectation value of the electron ensemble. The quantization axis z is determined by the direction of sample magnetization. Due to the electron spin being antiparallel to the magnetic moment in ferromagnets, the minority-character spin is referred to the sample magnetization direction [72].

By using of the well-defined oppositely polarized incident beams with changing the remanent magnetization direction of the Co films, we experimentally demonstrate the SPEELS spectral profiles strongly depend on the relative orientations of the incident beam polarization and the sample magnetization. In other words, the spectral weight of the magnetic-excitation peaks is dominated by the component of the polarized incident beams $\mathbf{P}$ onto the sample magnetization direction $\mathbf{M}$, i.e., incident electrons with the minority-spin character. Thus, the resulting asymmetry of the magnon peaks originating from the oppositely polarized incident beams is associated with the scalar product $\mathbf{M} \cdot \mathbf{P}$. By performing SPEELS measurements, one can qualitatively verify the direction of the remanent magnetization of the ferromagnetic films.

Accordingly, the polarization of the electron beam in measurements can be referred to the minority- and the majority-character spins (corresponding to the minority- and majority-spin defined in ferromagnets), which are pointing parallel and antiparallel to the sample magnetization direction, respectively. Here we assume a fully polarized incoming electron beam, i.e., $\mathbf{P_0} = 1$. As discussed above, the incident beam has an equal number of minority and majority electrons if the polarization vector of the incident beam is perpendicular to the sample magnetization direction ($\mathbf{P} \perp \mathbf{M}$), while it has only minority electrons if its polarization is parallel to the sample magnetization direction ($\mathbf{P} \parallel \mathbf{M}$).

In this study, the spin-polarized inelastic electron scattering measurements with different relative orientation (0°, 45°, 90°, 135°, 270°, and 180°) between the polarization vectors of the primary electron beam and the sample magnetization are demonstrated. It is shown that the inelastic spectral weight strongly depends on the relative orientation between the incident beam polarization and the sample magnetization (see Sec. 4.1.3 and 4.3.3).

Bibliography

[1] D. Grohol, K. Matan, J.-H. Cho, S.-H. Lee, J. W. Lynn, D. G. Nocera, and Y. S. Lee, "Spin chirality on a two-dimensional frustrated lattice," *Nat Mater*, vol. 4, pp. 323–328, Apr. 2005. 1

[2] I. Mirebeau and S. Petit, "Magnetic frustration probed by inelastic neutron scattering: Recent examples," *Journal of Magnetism and Magnetic Materials*, vol. 350, pp. 209 – 216, 2014. 1

[3] P. Dai, "Antiferromagnetic order and spin dynamics in iron-based superconductors," *Rev. Mod. Phys.*, vol. 87, pp. 855–896, Aug 2015. 1

[4] P. Monthoux and D. Pines, "Spin-fluctuation-induced superconductivity in the copper oxides: A strong coupling calculation," *Phys. Rev. Lett.*, vol. 69, pp. 961–964, Aug 1992. 1

[5] F. Essenberger, A. Sanna, A. Linscheid, F. Tandetzky, G. Profeta, P. Cudazzo, and E. K. U. Gross, "Superconducting pairing mediated by spin fluctuations from first principles," *Phys. Rev. B*, vol. 90, p. 214504, Dec 2014. 1

[6] J. Lischner, T. Bazhirov, A. H. MacDonald, M. L. Cohen, and S. G. Louie, "First-principles theory of electron-spin fluctuation coupling and superconducting instabilities in iron selenide," *Phys. Rev. B*, vol. 91, p. 020502, Jan 2015. 1

[7] J. Paglione and R. L. Greene, "High-temperature superconductivity in iron-based materials," *Nat Phys*, vol. 6, pp. 645–658, Sept. 2010. 1

[8] L. J. Cornelissen, J. Liu, R. A. Duine, J. B. Youssef, and B. J. van Wees, "Long-distance transport of magnon spin information in a magnetic insulator at room temperature," *Nat Phys*, vol. 11, pp. 1022–1026, Dec. 2015. 1

[9] A. V. Chumak, A. A. Serga, and B. Hillebrands, "Magnon transistor for all-magnon data processing," *Nat Commun*, vol. 5, Aug. 2014. 1

[10] K. N. Trohidou, J. A. Blackman, and J. F. Cooke, "Calculation of the high-energy spin-wave spectrum of hcp cobalt," *Phys. Rev. Lett.*, vol. 67, pp. 2561–2564, Oct 1991. 1

[11] T. Perring, A. Taylor, and G. Squires, "High-energy spin waves in hexagonal cobalt," *Physica B: Condensed Matter*, vol. 213-214, pp. 348–350, Aug 1995. 1

[12] J. Bass, J. Blackman, and J. Cooke, "The role of the exchange matrix in the itinerant-electron theory of ferromagnetism," *Journal of Physics: Condensed Matter*, vol. 4, no. 16, p. L275, 1992. 1

[13] P. Buczek, A. Ernst, and L. M. Sandratskii, "Interface electronic complexes and landau damping of magnons in ultrathin magnets," *Phys. Rev. Lett.*, vol. 106, p. 157204, Apr 2011. 1, 5.5.1

[14] E. Şaşıoğlu, A. Schindlmayr, C. Friedrich, F. Freimuth, and S. Blügel, "Wannier-function approach to spin excitations in solids," *Phys. Rev. B*, vol. 81, p. 054434, Feb 2010. 1, 5.4, 5.4

[15] P. Buczek, A. Ernst, and L. M. Sandratskii, "Different dimensionality trends in the landau damping of magnons in iron, cobalt, and nickel: Time-dependent density functional study," *Phys. Rev. B*, vol. 84, p. 174418, Nov 2011. 1, 4.3.3, 5.3, 5.5.1

[16] J. F. Cooke, J. A. Blackman, and T. Morgan, "New interpretation of spin-wave behavior in nickel," *Phys. Rev. Lett.*, vol. 54, pp. 718–721, Feb 1985. 1

[17] J. Sánchez-Barriga, J. Braun, J. Minár, I. Di Marco, A. Varykhalov, O. Rader, V. Boni, V. Bellini, F. Manghi, H. Ebert, M. I. Katsnelson, A. I. Lichtenstein, O. Eriksson, W. Eberhardt, H. A. Dürr, and J. Fink, "Effects of spin-dependent quasiparticle renormalization in Fe, Co, and Ni photoemission spectra:an experimental and theoretical study," *Phys. Rev. B*, vol. 85, p. 205109, May 2012. 1, 5.4, 8, 6

[18] S. Monastra, F. Manghi, C. A. Rozzi, C. Arcangeli, E. Wetli, H.-J. Neff, T. Greber, and J. Osterwalder, "Quenching of majority-channel quasiparticle excitations in cobalt," *Phys. Rev. Lett.*, vol. 88, p. 236402, May 2002. 1, 1, 5.4, 6

[19] J. Braun, J. Minár, H. Ebert, M. I. Katsnelson, and A. I. Lichtenstein, "Spectral function of ferromagnetic 3d metals: A self-consistent LSDA + DMFT approach combined with the one-step model of photoemission," *Phys. Rev. Lett.*, vol. 97, p. 227601, Dec 2006. 1, 5.4, 5.4

[20] J. Schäfer, D. Schrupp, E. Rotenberg, K. Rossnagel, H. Koh, P. Blaha, and R. Claessen, "Electronic quasiparticle renormalization on the spin wave energy scale," *Phys. Rev. Lett.*, vol. 92, p. 097205, Mar 2004. 1

[21] A. Hofmann, X. Y. Cui, J. Schäfer, S. Meyer, P. Höpfner, C. Blumenstein, M. Paul, L. Patthey, E. Rotenberg, J. Bünemann, F. Gebhard, T. Ohm, W. Weber, and R. Claessen, "Renormalization of bulk magnetic electron states at high binding energies," *Phys. Rev. Lett.*, vol. 102, p. 187204, May 2009. 1

[22] A. Freeman and R. quian Wu, "Electronic structure theory of surface, interface and thin-film magnetism," *Journal of Magnetism and Magnetic Materials*, vol. 100, no. 1, pp. 497 – 514, 1991. 1, 5.1.1, 5.1.2, 5.3

[23] M. Tischer, O. Hjortstam, D. Arvanitis, J. Hunter Dunn, F. May, K. Baberschke, J. Trygg, J. M. Wills, B. Johansson, and O. Eriksson, "Enhancement of orbital magnetism at surfaces: Co on Cu(100)," *Phys. Rev. Lett.*, vol. 75, pp. 1602–1605, Aug 1995. 1, 5.1.2

[24] D.-s. Wang, A. J. Freeman, and H. Krakauer, "Surface magnetism of a Ni overlayer on a Cu(001) substrate," *Phys. Rev. B*, vol. 24, pp. 1126–1129, Jul 1981. 1, 5.1.2

[25] X. Qian and W. Hübner, "First-principles calculation of structural and magnetic properties for Fe monolayers and bilayers on W(110)," *Phys. Rev. B*, vol. 60, pp. 16192–16197, Dec 1999. 1, 5.1.2

[26] C. F. Hirjibehedin, C. P. Lutz, and A. J. Heinrich, "Spin coupling in engineered atomic structures," *Science*, vol. 312, no. 5776, pp. 1021–1024, 2006. 1

[27] T. Balashov, P. Buczek, L. Sandratskii, A. Ernst, and W. Wulfhekel, "Magnon dispersion in thin magnetic films," *Journal of Physics: Condensed Matter*, vol. 26, no. 39, p. 394007, 2014. 1

[28] A. Spinelli, B. Bryant, F. Delgado, J. Fernández-Rossier, and A. Otte, "Imaging of spin waves in atomically designed nanomagnets," *Nature materials*, vol. 13, no. 8, pp. 782–785, 2014. 1

[29] R. Vollmer, M. Etzkorn, P. S. A. Kumar, H. Ibach, and J. Kirschner, "Spin-polarized electron energy loss spectroscopy of high energy, large wave vector spin waves in ultrathin fcc Co films on Cu(001)," *Phys. Rev. Lett.*, vol. 91, p. 147201, Sep 2003. 1, 3.1.2

[30] T. Balashov, A. Takács, W. Wulfhekel, and J. Kirschner, "Magnon excitation with spin-polarized scanning tunneling microscopy," *Physical review letters*, vol. 97, no. 18, p. 187201, 2006. 1

[31] H. Ibach, D. Bruchmann, R. Vollmer, M. Etzkorn, P. A. Kumar, and J. Kirschner, "A novel spectrometer for spin-polarized electron energy-loss spectroscopy," *Review of scientific instruments*, vol. 74, no. 9, pp. 4089–4095, 2003. 1, 3.2.2

[32] J. Sánchez-Barriga, J. Minár, J. Braun, A. Varykhalov, V. Boni, I. Di Marco, O. Rader, V. Bellini, F. Manghi, H. Ebert, M. I. Katsnelson, A. I. Lichtenstein, O. Eriksson, W. Eberhardt, H. A. Dürr, and J. Fink, "Quantitative determination of spin-dependent quasiparticle lifetimes and electronic correlations in hcp cobalt," *Phys. Rev. B*, vol. 82, p. 104414, Sep 2010. 1, 5.4, 6

[33] P. Mohn, *Magnetism in the solid state: an introduction*, vol. 134. Springer Science & Business Media, 2006. 2

[34] M. Getzlaff, J. Bansmann, and G. Schönhense, "Spin-polarization effects for electrons passing through thin iron and cobalt films," *Solid state communications*, vol. 87, no. 5, pp. 467–469, 1993. 2.3.1

[35] E. Vescovo, C. Carbone, U. Alkemper, O. Rader, T. Kachel, W. Gudat, and W. Eberhardt, "Spin-dependent electron scattering in ferromagnetic Co layers on Cu(111)," *Phys. Rev. B*, vol. 52, pp. 13497–13503, Nov 1995. 2.3.1

[36] J. W. Gadzuk, *Vibrational Spectroscopy of Molecules on Surfaces*, ch. Excitation Mechanisms in Vibrational Spectroscopy of Molecules on Surfaces, pp. 49–103. Boston, MA: Springer US, 1987. 2.3.1

[37] H. Ibach and D. L. Mills, *Electron energy loss spectroscopy and surface vibrations*. Academic Press, 1982. 2.3.1

[38] H. Ibach, *Electron energy loss spectrometers*. Springer, 1991. 2.3.1

[39] J. Kessler, *Polarized electrons*. Springer Berlin Heidelberg, 1985. 2.3.2

[40] J. Krischner, "Polarized electrons at surfaces," *Springer Tracts in Modern Physics*, 1985. 2.3.2, 3, 3.1.1

[41] J. Kirschner, "Direct and exchange contributions in inelastic scattering of spin-polarized electrons from iron," *Phys. Rev. Lett.*, vol. 55, pp. 973–976, Aug 1985. 3.1.1

[42] D. Venus and J. Kirschner, "Momentum dependence of the stoner excitation spectrum of iron using spin-polarized electron-energy-loss spectroscopy," *Phys. Rev. B*, vol. 37, pp. 2199–2211, Feb 1988. 3.1.1

[43] G. Vignale and K. S. Singwi, "Spin-flip electron-energy-loss spectroscopy in itinerant-electron ferromagnets: Collective modes versus stoner excitations," *Phys. Rev. B*, vol. 32, pp. 2824–2834, Sep 1985. 3.1.1

[44] K.-P. Kämper, D. L. Abraham, and H. Hopster, "Spin-polarized electron-energy-loss spectroscopy on epitaxial fcc Co layers on Cu(001)," *Phys. Rev. B*, vol. 45, pp. 14335–14346, Jun 1992. 3.1.2

[45] M. Plihal, D. L. Mills, and J. Kirschner, "Spin wave signature in the spin polarized electron energy loss spectrum of ultrathin Fe films: Theory and experiment," *Phys. Rev. Lett.*, vol. 82, pp. 2579–2582, Mar 1999. 3.1.2

[46] T.-H. Chuang, *High wave-vector magnon excitations in ultrathin Fe(111) films grown on Au/W(110) and Fe(001) films grown on Ir(001)*. Ph.D. thesis, Martin-Luther University Halle-Wittenberg, Halle, 2013. 3.5

[47] D. T. Pierce, R. J. Celotta, G.-C. Wang, W. N. Unertl, A. Galejs, C. E. Kuyatt, and S. R. Mielczarek, "The GaAs spin polarized electron source," *Review of Scientific Instruments*, vol. 51, no. 4, pp. 478–499, 1980. 3.2.3

[48] P. Drescher, H. Andresen, K. Aulenbacher, J. Bermuth, T. Dombol, H. Fischerz, H. Euteneuer, N. Faleev, M. Galaktionov, D. v. Harrach, *et al.*, "Photoemission of spinpolarized electrons from strained GaAsP," *Applied Physics A*, vol. 63, no. 2, pp. 203–206, 1996. 3.2.3

[49] T. Nakanishi, H. Aoyagi, H. Horinaka, Y. Kamiya, T. Kato, S. Nakamura, T. Saka, and M. Tsubata, "Large enhancement of spin polarization observed by photoelectrons from a strained GaAs layer," *Physics Letters A*, vol. 158, no. 6, pp. 345 – 349, 1991. 3.2.3

[50] D. T. Pierce and F. Meier, "Photoemission of spin-polarized electrons from GaAs," *Phys. Rev. B*, vol. 13, pp. 5484–5500, Jun 1976. 3.2.3

[51] J. Kirschner, H. P. Oepen, and H. Ibach, "Energy- and spin-analysis of polarized photoelectrons from NEA GaAsP," *Applied Physics A: Materials Science & Processing*, vol. 30, pp. 177–183, 1983. 10.1007/BF00620537. 3.2.3

[52] T.-H. Chuang, K. Zakeri, A. Ernst, Y. Zhang, H. Qin, Y. Meng, Y.-J. Chen, and J. Kirschner, "Magnetic properties and magnon excitations in Fe (001)

films grown on Ir (001)," *Physical Review B*, vol. 89, no. 17, p. 174404, 2014. 4.1.1, 5.1.1

[53] K. Heinz and L. Hammer, "Combined application of LEED and STM in surface crystallography," *The Journal of Physical Chemistry B*, vol. 108, no. 38, pp. 14579–14584, 2004. 4.1.1, 4.1

[54] J. Kirschner, H. Engelhard, and D. Hartung, "An evaporation source for ion beam assisted deposition in ultrahigh vacuum," *Review of scientific instruments*, vol. 73, no. 11, pp. 3853–3860, 2002. 4.1.1

[55] A. Klein, A. Schmidt, W. Meyer, L. Hammer, and K. Heinz, "Growth of metal nanowires of tunable width," *Phys. Rev. B*, vol. 81, p. 115431, Mar 2010. 4.1.1, 1, 4.1.2, 4.4, 5.3

[56] C. Giovanardi, A. Klein, A. Schmidt, L. Hammer, and K. Heinz, "Transition metal superlattices and epitaxial films on Ir(100)-(5 × 1)-H," *Phys. Rev. B*, vol. 78, p. 205416, Nov 2008. 1

[57] L. Hammer, W. Meier, A. Schmidt, and K. Heinz, "Submonolayer iron film growth on reconstructed Ir(100) − (5 × 1)," *Phys. Rev. B*, vol. 67, p. 125422, Mar 2003. 4.1.1

[58] K. Heinz and L. Hammer, "Nanostructure formation on Ir(100)," *Progress in Surface Science*, vol. 84, no. 1, pp. 2–17, 2009. 4.1.1, 4.1.2, 4.4, 5.3

[59] G. Gilarowski, J. Méndez, and H. Niehus, "Initial growth of Cu on Ir(100)-(5× 1)," *Surface Science*, vol. 448, no. 2, pp. 290–304, 2000. 4.1.1

[60] M. Gubo, L. Hammer, and K. Heinz, "Laterally nanostructured cobalt oxide films on Ir(100)," *Phys. Rev. B*, vol. 85, p. 113402, Mar 2012. 4.1.2, 4.4

[61] C. M. Schneider, P. Bressler, P. Schuster, J. Kirschner, J. J. de Miguel, and R. Miranda, "Curie temperature of ultrathin films of fcc-cobalt epitaxially grown on atomically flat Cu(100) surfaces," *Phys. Rev. Lett.*, vol. 64, pp. 1059–1062, Feb 1990. 4.1.2

[62] W. Weber, C. H. Back, A. Bischof, C. Würsch, and R. Allenspach, "Morphology-induced oscillations of the magnetic anisotropy in ultrathin Co films," *Phys. Rev. Lett.*, vol. 76, pp. 1940–1943, Mar 1996. 4.1.2

[63] P. Gambardella, A. Dallmeyer, K. Maiti, M. Malagoli, W. Eberhardt, K. Kern, and C. Carbone, "Ferromagnetism in one-dimensional monatomic metal chains," *Nature*, vol. 416, no. 6878, pp. 301–304, 2002. 4.1.2

[64] B. Dupé, J. E. Bickel, Y. Mokrousov, F. Otte, K. von Bergmann, A. Kubetzka, S. Heinze, and R. Wiesendanger, "Giant magnetization canting due to symmetry breaking in zigzag Co chains on Ir (001)," *New Journal of Physics*, vol. 17, no. 2, p. 023014, 2015. 4.1.2

[65] N. Nakajima, T. Koide, T. Shidara, H. Miyauchi, H. Fukutani, A. Fujimori, K. Iio, T. Katayama, M. Nývlt, and Y. Suzuki, "Perpendicular magnetic anisotropy caused by interfacial hybridization via enhanced orbital moment in Co/Pt multilayers: Magnetic circular x-ray dichroism study," *Phys. Rev. Lett.*, vol. 81, pp. 5229–5232, Dec 1998. 4.1.2, 4.3.2

[66] S. Bornemann, O. Šipr, S. Mankovsky, S. Polesya, J. B. Staunton, W. Wurth, H. Ebert, and J. Minár, "Trends in the magnetic properties of Fe, Co, and Ni clusters and monolayers on Ir(111), Pt(111), and Au(111)," *Phys. Rev. B*, vol. 86, p. 104436, Sep 2012. 4.1.2, 5.3

[67] D. sheng Wang, R. Wu, and A. Freeman, "Magnetocrystalline anisotropy of interfaces: first-principles theory for Co-Cu interface and interpretation by an effective ligand interaction model," *Journal of Magnetism and Magnetic Materials*, vol. 129, no. 2, pp. 237 – 258, 1994. 4.1.2

[68] F. Den Broeder, W. Hoving, and P. Bloemen, "Magnetic anisotropy of multilayers," *Journal of magnetism and magnetic materials*, vol. 93, pp. 562–570, 1991. 4.1.2

[69] G. H. O. Daalderop, P. J. Kelly, and M. F. H. Schuurmans, "Magnetic anisotropy of a free-standing Co monolayer and of multilayers which contain Co monolayers," *Phys. Rev. B*, vol. 50, pp. 9989–10003, Oct 1994. 4.1.2

[70] J. Zak, E. Moog, C. Liu, and S. Bader, "Additivity of the kerr effect in thin-film magnetic systems," *Journal of Magnetism and Magnetic Materials*, vol. 88, no. 3, pp. L261 – L266, 1990. 4.1.2

[71] Z. Q. Qiu, J. Pearson, and S. D. Bader, "Magneto-optic kerr ellipticity of epitaxial Co/Cu overlayers and superlattices," *Phys. Rev. B*, vol. 46, pp. 8195–8200, Oct 1992. 4.1.2

[72] K. B. Hathaway, *Ultrathin Magnetic Structures II: Measurement Techniques and Novel Magnetic Properties*, ch. Magnetic Coupling and Magnetoresistance, pp. 45–194. Berlin, Heidelberg: Springer Berlin Heidelberg, 1994. 4.1.3, A

[73] M. Etzkorn, *Spin waves with high energy and momentum in ultrathin Co-films studied by spin-polarized electron energy loss spectroscopy*. Ph.D. thesis, Martin-Luther University Halle-Wittenberg, Halle, 2005. 4.2, 4.13

[74] M. P. Gokhale, A. Ormeci, and D. L. Mills, "Inelastic scattering of low-energy electrons by spin excitations on ferromagnets," *Phys. Rev. B*, vol. 46, pp. 8978–8993, Oct 1992. 3

[75] E. Lundgren, G. Leonardelli, M. Schmid, and P. Varga, "A misfit structure in the Co/Pt (111) system studied by scanning tunnelling microscopy and embedded atom method calculations," *Surface Science*, vol. 498, no. 3, pp. 257–265, 2002. 4.3.1

[76] P. Grütter and U. T. Dürig, "Growth of vapor-deposited cobalt films on Pt(111) studied by scanning tunneling microscopy," *Phys. Rev. B*, vol. 49, pp. 2021–2029, Jan 1994. 4.3.1, 4.3.2

[77] E. Lundgren, B. Stanka, M. Schmid, and P. Varga, "Thin films of Co on Pt(111): Strain relaxation and growth," *Phys. Rev. B*, vol. 62, pp. 2843–2851, Jul 2000. 4.3.1, 4.3.2, 5.1.1

[78] R. Baudoing-Savois, P. Dolle, Y. Gauthier, M. Saint-Lager, M. De Santis, and V. Jahns, "Co ultra-thin films on Pt (111) and Co-Pt alloying: a LEED, Auger and synchrotron x-ray diffraction study," *Journal of Physics: Condensed Matter*, vol. 11, no. 43, p. 8355, 1999. 4.3.1

[79] A. Lehnert, S. Dennler, P. Błoński, S. Rusponi, M. Etzkorn, G. Moulas, P. Bencok, P. Gambardella, H. Brune, and J. Hafner, "Magnetic anisotropy of Fe and Co ultrathin films deposited on Rh(111) and Pt(111) substrates: An experimental and first-principles investigation," *Phys. Rev. B*, vol. 82, p. 094409, Sep 2010. 4.3.2, 5.11

[80] G. Moulas, A. Lehnert, S. Rusponi, J. Zabloudil, C. Etz, S. Ouazi, M. Etzkorn, P. Bencok, P. Gambardella, P. Weinberger, and H. Brune, "High magnetic moments and anisotropies for Fe_xCo_{1-x} monolayers on Pt(111)," *Phys. Rev. B*, vol. 78, p. 214424, Dec 2008. 4.3.2

[81] P. Grütter and U. Dürig, "Scanning tunneling microscopy of Co on Pt (111)," *Journal of Vacuum Science & Technology B*, vol. 12, no. 3, pp. 1768–1771, 1994. 4.3.2, 5.1.1

[82] U. Pustogowa, J. Zabloudil, C. Uiberacker, C. Blaas, P. Weinberger, L. Szunyogh, and C. Sommers, "Magnetic properties of thin films of Co and of (CoPt) superstructures on Pt(100) and Pt(111)," *Phys. Rev. B*, vol. 60, pp. 414–421, Jul 1999. 4.3.2

[83] R. Allenspach, "Ultrathin films: magnetism on the microscopic scale," *Journal of Magnetism and Magnetic Materials*, vol. 129, no. 2, pp. 160–185, 1994. 4.3.2

[84] H. Watanabe, K. Amemiya, D. Matsumura, S. Kitagawa, H. Abe, T. Yokoyama, and T. Ohta, "XMCD study of spin reorientation transitions of Co/Pt(111) induced by CO adsorption," *Photon Factory Activity Report 2002 Part B*, vol. 20, p. 61, 2003. 4.3.2

[85] A. T. Costa, R. B. Muniz, and D. L. Mills, "Theory of spin waves in ultrathin ferromagnetic films: The case of Co on Cu(100)," *Phys. Rev. B*, vol. 69, p. 064413, Feb 2004. 4.3.3

[86] H. Fritzsche, J. Kohlhepp, and U. Gradmann, "Epitaxial strain and magnetic anisotropy in ultrathin Co films on W(110)," *Phys. Rev. B*, vol. 51, pp. 15933–15941, Jun 1995. 5.1

[87] M. Pratzer, H. J. Elmers, and M. Getzlaff, "Heteroepitaxial growth of Co on W(110) investigated by scanning tunneling microscopy," *Phys. Rev. B*, vol. 67, p. 153405, Apr 2003. 5.1.1

[88] M. Pratzer and H. J. Elmers, "Scanning tunneling spectroscopy of dislocations in ultrathin fcc and hcp Co films," *Phys. Rev. B*, vol. 72, p. 035460, Jul 2005. 5.1.1

[89] J. Rajeswari, H. Ibach, and C. M. Schneider, "Standing spin waves in ultrathin magnetic films: A method to test for layer-dependent exchange coupling," *Phys. Rev. Lett.*, vol. 112, p. 127202, Mar 2014. 5.1.1

[90] E. Michel, H. Ibach, and C. M. Schneider, "Spin waves in ultrathin hexagonal cobalt films on W(110), Cu(111), and Au(111) surfaces," *Phys. Rev. B*, vol. 92, p. 024407, Jul 2015. 5.1.1

[91] H. J. Qin, K. Zakeri, A. Ernst, T.-H. Chuang, Y.-J. Chen, Y. Meng, and J. Kirschner, "Magnons in ultrathin ferromagnetic films with a large perpendicular magnetic anisotropy," *Phys. Rev. B*, vol. 88, p. 020404, Jul 2013. 5.1.1

[92] J. Rajeswari, H. Ibach, and C. M. Schneider, "Large wave vector surface spin waves of the nanomartensitic phase in ultrathin iron films on Cu(100)," *EPL (Europhysics Letters)*, vol. 101, no. 1, p. 17003, 2013. 5.1.1

[93] Y. Meng, K. Zakeri, A. Ernst, T.-H. Chuang, H. J. Qin, Y.-J. Chen, and J. Kirschner, "Direct evidence of antiferromagnetic exchange interaction in Fe(001) films: Strong magnon softening at the high-symmetry $\overline{\mathrm{M}}$ point," *Phys. Rev. B*, vol. 90, p. 174437, Nov 2014. 5.1.1

[94] H. Qin, K. Zakeri, A. Ernst, L. Sandratskii, P. Buczek, A. Marmodoro, T.-H. Chuang, Y. Zhang, and J. Kirschner, "Long-living terahertz magnons in

ultrathin metallic ferromagnets," *Nature communications*, vol. 6, 2015. 5.1.1, 5.5.1, 5.5.1

[95] J. Prokop, W. X. Tang, Y. Zhang, I. Tudosa, T. R. F. Peixoto, K. Zakeri, and J. Kirschner, "Magnons in a ferromagnetic monolayer," *Phys. Rev. Lett.*, vol. 102, p. 177206, Apr 2009. 5.1.1, 5.5.1

[96] M. Pajda, J. Kudrnovský, I. Turek, V. Drchal, and P. Bruno, "*Ab initio* calculations of exchange interactions, spin-wave stiffness constants, and curie temperatures of Fe, Co, and Ni," *Phys. Rev. B*, vol. 64, p. 174402, Oct 2001. 5.1.1

[97] M. Pajda, J. Kudrnovský, I. Turek, V. Drchal, and P. Bruno, "Oscillatory curie temperature of two-dimensional ferromagnets," *Phys. Rev. Lett.*, vol. 85, pp. 5424–5427, Dec 2000. 5.1.1

[98] B. Rousseau, A. Eiguren, and A. Bergara, "Efficient computation of magnon dispersions within time-dependent density functional theory using maximally localized wannier functions," *Phys. Rev. B*, vol. 85, p. 054305, Feb 2012. 5.2

[99] E. C. Bain and N. Dunkirk, "The nature of martensite," *trans. AIME*, vol. 70, no. 1, pp. 25–47, 1924. 3

[100] J. P. Perdew, K. Burke, and M. Ernzerhof, "Generalized gradient approximation made simple," *Phys. Rev. Lett.*, vol. 77, pp. 3865–3868, Oct 1996. 5.3

[101] M. Lüders, A. Ernst, W. M. Temmerman, Z. Szotek, and P. J. Durham, "*Ab initio* angle-resolved photoemission in multiple-scattering formulation," *Journal of Physics: Condensed Matter*, vol. 13, no. 38, p. 8587, 2001. 5.3

[102] A. Liechtenstein, M. Katsnelson, V. Antropov, and V. Gubanov, "Local spin density functional approach to the theory of exchange interactions in ferromagnetic metals and alloys," *Journal of Magnetism and Magnetic Materials*, vol. 67, no. 1, pp. 65 – 74, 1987. 5.3

[103] T.-H. Chuang, K. Zakeri, A. Ernst, L. M. Sandratskii, P. Buczek, Y. Zhang, H. J. Qin, W. Adeagbo, W. Hergert, and J. Kirschner, "Impact of atomic structure on the magnon dispersion relation: A comparison between Fe(111)/Au/W(110) and Fe(110)/W(110)," *Phys. Rev. Lett.*, vol. 109, p. 207201, Nov 2012. 5.3

[104] D. Sander, "The magnetic anisotropy and spin reorientation of nanostructures and nanoscale films," *Journal of Physics: Condensed Matter*, vol. 16, no. 20, p. R603, 2004. 5.3

[105] C. Etz, L. Bergqvist, A. Bergman, A. Taroni, and O. Eriksson, "Atomistic spin dynamics and surface magnons," *Journal of Physics: Condensed Matter*, vol. 27, no. 24, p. 243202, 2015. 5.3

[106] P. Gambardella, S. Rusponi, M. Veronese, S. S. Dhesi, C. Grazioli, A. Dallmeyer, I. Cabria, R. Zeller, P. H. Dederichs, K. Kern, C. Carbone, and H. Brune, "Giant magnetic anisotropy of single cobalt atoms and nanoparticles," *Science*, vol. 300, no. 5622, pp. 1130–1133, 2003. 5.3

[107] C. Etz, J. Zabloudil, P. Weinberger, and E. Y. Vedmedenko, "Magnetic properties of single atoms of Fe and Co on Ir(111) and Pt(111)," *Phys. Rev. B*, vol. 77, p. 184425, May 2008. 5.3

[108] J. E. Bickel, F. Meier, J. Brede, A. Kubetzka, K. von Bergmann, and R. Wiesendanger, "Magnetic properties of monolayer Co islands on Ir(111) probed by spin-resolved scanning tunneling microscopy," *Phys. Rev. B*, vol. 84, p. 054454, Aug 2011. 5.3

[109] D. Wang, R. Wu, and A. J. Freeman, "Local spin density theory of interface and surface magnetocrystalline anisotropy: Pd/Co/Pd(001) and Cu/Co/Cu(001) sandwiches," *Journal of Applied Physics*, vol. 75, no. 10, pp. 6409–6411, 1994. 5.3

[110] A. Taroni, A. Bergman, L. Bergqvist, J. Hellsvik, and O. Eriksson, "Suppression of standing spin waves in low-dimensional ferromagnets," *Phys. Rev. Lett.*, vol. 107, p. 037202, Jul 2011. 5.4

[111] M. M. Steiner, R. C. Albers, and L. J. Sham, "Quasiparticle properties of Fe, Co, and Ni," *Phys. Rev. B*, vol. 45, pp. 13272–13284, Jun 1992. 5.4

[112] D. M. Edwards and R. B. Muniz, "Spin waves in ferromagnetic transition metals. i. general formalism and application to nickel and its alloys," *Journal of Physics F: Metal Physics*, vol. 15, no. 11, p. 2339, 1985. 5.4

[113] R. B. Muniz, J. F. Cooke, and D. M. Edwards, "Spin waves in ferromagnetic transition metals. ii. iron and its alloys," *Journal of Physics F: Metal Physics*, vol. 15, no. 11, p. 2357, 1985. 5.4

[114] K. Karlsson and F. Aryasetiawan, "Spin-wave excitation spectra of nickel and iron," *Phys. Rev. B*, vol. 62, pp. 3006–3009, Aug 2000. 5.4

[115] A. T. Costa, R. B. Muniz, and D. L. Mills, "Theory of large-wave-vector spin waves in ultrathin ferromagnetic films: Sensitivity to electronic structure," *Phys. Rev. B*, vol. 70, p. 054406, Aug 2004. 5.4

[116] M. Mulazzi, J. Miyawaki, A. Chainani, Y. Takata, M. Taguchi, M. Oura, Y. Senba, H. Ohashi, and S. Shin, "Fermi surface of Co(0001) and initial-state linewidths determined by soft x-ray angle-resolved photoemission spectroscopy," *Phys. Rev. B*, vol. 80, p. 241106, Dec 2009. 5.4

[117] Y. Zhang, T.-H. Chuang, K. Zakeri, and J. Kirschner, "Relaxation time of terahertz magnons excited at ferromagnetic surfaces," *Phys. Rev. Lett.*, vol. 109, p. 087203, Aug 2012. 5.5.1

[118] W. X. Tang, Y. Zhang, I. Tudosa, J. Prokop, M. Etzkorn, and J. Kirschner, "Large wave vector spin waves and dispersion in two monolayer Fe on $W(110)$," *Phys. Rev. Lett.*, vol. 99, p. 087202, Aug 2007. 5.5.1

[119] A. Khitun, "Multi-frequency magnonic logic circuits for parallel data processing," *Journal of Applied Physics*, vol. 111, no. 5, 2012. 6

[120] A. Khitun, M. Bao, and K. L. Wang, "Magnonic logic circuits," *Journal of Physics D: Applied Physics*, vol. 43, no. 26, p. 264005, 2010. 6

[121] X. Xing, Q. Jin, and S. Li, "Frequency-selective manipulation of spin waves: micromagnetic texture as amplitude valve and mode modulator," *New Journal of Physics*, vol. 17, no. 2, p. 023020, 2015. 6

[122] A. Khitun, "Magnonic holographic devices for special type data processing," *Journal of Applied Physics*, vol. 113, no. 16, 2013. 6

[123] A. Khitun and K. L. Wang, "Non-volatile magnonic logic circuits engineering," *Journal of Applied Physics*, vol. 110, no. 3, 2011. 6

[124] K.-S. Lee, D.-S. Han, and S.-K. Kim, "Physical origin and generic control of magnonic band gaps of dipole-exchange spin waves in width-modulated nanostrip waveguides," *Phys. Rev. Lett.*, vol. 102, p. 127202, Mar 2009. 6

[125] M. I. Makin, J. H. Cole, C. D. Hill, and A. D. Greentree, "Spin guides and spin splitters: Waveguide analogies in one-dimensional spin chains," *Phys. Rev. Lett.*, vol. 108, p. 017207, Jan 2012. 6

[126] J. Linder and J. W. A. Robinson, "Superconducting spintronics," *Nat Phys*, vol. 11, pp. 307–315, Apr. 2015. 6

[127] S. Tan, Y. Zhang, M. Xia, Z. Ye, F. Chen, X. Xie, R. Peng, D. Xu, Q. Fan, H. Xu, J. Jiang, T. Zhang, X. Lai, T. Xiang, J. Hu, B. Xie, and D. Feng, "Interface-induced superconductivity and strain-dependent spin density waves in $FeSe/SrTiO_3$ thin films," *Nat Mater*, vol. 12, pp. 634–640, July 2013. 6

[128] D. Stornaiuolo, C. Cantoni, G. M. De Luca, R. Di Capua, E. Di. Gennaro, G. Ghiringhelli, B. Jouault, D. Marre, D. Massarotti, F. Miletto Granozio, I. Pallecchi, C. Piamonteze, S. Rusponi, F. Tafuri, and M. Salluzzo, "Tunable spin polarization and superconductivity in engineered oxide interfaces," *Nat Mater*, vol. 15, pp. 278–283, Mar. 2016. 6

[129] M.-Y. Li, Y. Shi, C.-C. Cheng, L.-S. Lu, Y.-C. Lin, H.-L. Tang, M.-L. Tsai, C.-W. Chu, K.-H. Wei, J.-H. He, W.-H. Chang, K. Suenaga, and L.-J. Li, "Epitaxial growth of a monolayer WSe_2-MoS_2 lateral p-n junction with an atomically sharp interface," *Science*, vol. 349, no. 6247, pp. 524–528, 2015. 6

Acknowledgment

I would like to sincerely thank Prof. *Jürgen Kirschner* for giving me the opportunity to carry out my doctoral research in his group at the Max-Planck-Institute of Microstructure Physics. I am deeply grateful that he always provides valuable suggestions, warm encouragement, and kind support during my Ph.D. years. The opportunity he gave me to present the work in national and international conferences and workshops is greatly appreciated.

I would like to express my gratitude to Dr. *Khalil Zakeri Lori* for introducing the field of magnetic excitations and spin-polarized electron energy-loss spectroscopy. I am also grateful for his valuable guidance and his continuous encouragement throughout my Ph.D. studies.

I would like to address my special thanks to Dr. *Christian Tusche*, who had generously provided me insightful comments during the course of my work. I am very grateful for his availability for discussions and his precious knowledge of many aspects of physics which I find considerably inspired.

I would like to sincerely thank Prof. *Stuart Parkin* for his helpful comments and discussions in this study and his support at the final stage of my PhD work.

I would like to thank members of the SPEELS group, Dr. *Tzu-Hung Chuang*, Dr. *huajun Qin,* and Dr. *Yang Meng*, who helped me getting started in performing experiments and data analysis.

I would like to thanks collaborators from the theory department, Dr. *Arthur Ernst,* Dr. *Leonid M. Sandratskii,* and Dr. *Paweł Buczek*. I especially thank Dr. *Arthur Ernst* for a fruitful collaboration on the theoretical aspects of this work.

I would like to acknowledge the technical support from Mr. *Herbert Engelhard,* Ms. *Heike Menge*, and the co-workers from the mechanical and electronics workshop.

I would also like to express my gratitude towards Ms. *Doreen Röder,* Ms. *Ina Goffin,* Ms. *Antje Paetzold,* Ms. *Simone Jäger,* and the co-workers in the administration office, Ms. *Birgit Frankenstein,* Ms. *Angelika Schneider,* and Ms. *Martina Witzig,* for their kind assistance in several administrative manners.

During my PhD, I have shared many memorable moments in Halle with colleges in the institute: Dr. *Martin Ellguth,* Dr. *Matthias Schmidt,* Dr. *Michael Huth,* Dr. *Ping Yu,* Dr. *Cheng-Tien Chiang,* Dr. *Zheng Wei,* Dr. *Yuichiro Matsushita,* Dr. *Yasushi Shinohara,* Mr. *Frank Helbig,* Mr. *Frank Weiß,* Mr. *Florian Thiele,* Ms. *Kenia Novakoski Fischer,* and Dr. *Jeison Antonio Fischer.* I also thank all the members in the *International Max Planck Research School* for sharing enjoyable scientific activities.

I would like to thanks to my great friends in Germany, *Bi-Lin Ho, Jia Xu, Ying Zhang, Chang-Yang Kuo, Meng-Jie Huang,* and *Yu-Chun Chen,* for their friendship and encouragement during my studies. I also deeply appreciate *Lo Yueh Chang* for his continued support and warm company throughout the years.

My heartfelt thanks and love are extended to my parents and the sister. Their love, support, happiness, and encouragement are the source of my strength to face challenges and to achieve goals in my life.

Publications

[1] H.-J. Liu, Y.-Y. Liu, C.-Y. Tsai, S.-C. Liao, **Y.-J. Chen**, H.-J. Lin, C.-H. Lai, W.-F. Hsieh, J.-Y. Li, C.-T. Chen, Q. He, and Y.-H. Chu, "Tuning the functionalities of a mesocrystal via structural coupling," *Scientific Reports*, vol. 5, July 2015.

[2] T.-H. Chuang, K. Zakeri, A. Ernst, Y. Zhang, H. J. Qin, Y. Meng, **Y.-J. Chen**, and J. Kirschner, "Magnetic properties and magnon excitations in Fe(001) films grown on Ir(001)," *Phys. Rev. B*, vol. 89, p. 174404, May 2014.

[3] Y. Meng, K. Zakeri, A. Ernst, T.-H. Chuang, H. J. Qin, **Y.-J. Chen**, and J. Kirschner, "Direct evidence of antiferromagnetic exchange interaction in Fe(001) films: Strong magnon softening at the high-symmetry $\overline{\text{M}}$ point," *Phys. Rev. B*, vol. 90, p. 174437, Nov 2014.

[4] H. J. Qin, K. Zakeri, A. Ernst, T.-H. Chuang, **Y.-J. Chen**, Y. Meng, and J. Kirschner, "Magnons in ultrathin ferromagnetic films with a large perpendicular magnetic anisotropy," *Phys. Rev. B*, vol. 88, p. 020404, Jul 2013.

[5] **Y.-J. Chen**, Y.-H. Hsieh, S.-C. Liao, Z. Hu, M.-J. Huang, W.-C. Kuo, Y.-Y. Chin, T.-M. Uen, J.-Y. Juang, C.-H. Lai, H.-J. Lin, C.-T. Chen, and Y.-H. Chu, "Strong magnetic enhancement in self-assembled multiferroic-ferrimagnetic nanostructures," *Nanoscale*, vol. 5, pp. 4449–4453, 2013.

[6] H.-J. Liu, V.-T. Tra, **Y.-J. Chen**, R. Huang, C.-G. Duan, Y.-H. Hsieh, H.-J. Lin, J.-Y. Lin, C.-T. Chen, Y. Ikuhara, and Y.-H. Chu, "Large magnetoresistance in magnetically coupled SrRuO$_3$-CoFe$_2$O$_4$ self-assembled nanostructures," *Advanced Materials*, vol. 25, no. 34, pp. 4753–4759, 2013.

[7] J.-L. Guo, Y.-D. Chiou, W.-I. Liang, H.-J. Liu, **Y.-J. Chen**, W.-C. Kuo, C.-Y. Tsai, K.-A. Tsai, H.-H. Kuo, W.-F. Hsieh, J.-Y. Juang, Y.-J. Hsu, H.-J. Lin, C.-T. Chen, X.-P. Liao, B. Shi, and Y.-H. Chu, "Complex oxide-noble metal

conjugated nanoparticles," *Advanced Materials*, vol. 25, no. 14, pp. 2040–2044, 2013.

[8] Y.-H. Hsieh, H.-H. Kuo, S.-C. Liao, H.-J. Liu, **Y.-J. Chen**, H.-J. Lin, C.-T. Chen, C.-H. Lai, Q. Zhan, Y.-L. Chueh, and Y.-H. Chu, "Tuning the formation and functionalities of ultrafine $CoFe_2O_4$ nanocrystals via interfacial coherent strain," *Nanoscale*, vol. 5, pp. 6219–6223, 2013.

[9] J. Cheung, K. Bogle, X. Cheng, J. Sullaphen, C.-Y. Kuo, **Y.-J. Chen**, H.-J. Lin, C.-T. Chen, J.-C. Yang, Y.-H. Chu, and N. Valanoor, "Phase evolution of magnetite nanocrystals on oxide supports via template-free bismuth ferrite precursor approach," *Journal of Applied Physics*, vol. 112, no. 10, 2012.

[10] **Y.-J. Chen**, C.-J. Hsu, C.-N. Liao, H.-T. Huang, C.-P. Lee, Y.-H. Chiu, T.-Y. Tung, and M.-F. Lai, "Controllable magnetization processes induced by nucleation sites in permalloy rings," *Japanese Journal of Applied Physics*, vol. 49, no. 2R, p. 023001, 2010.

[11] M.-F. Lai, C.-J. Hsu, C.-N. Liao, **Y.-J. Chen**, and Z.-H. Wei, "Magnetoresistance measurement of permalloy thin film rings with triangular fins," *Journal of Magnetism and Magnetic Materials*, vol. 322, no. 1, pp. 92 – 96, 2010.

[12] M. F. Lai, **Y.-J. Chen**, D. R. Liu, C. K. Lo, C. J. Hsu, C. N. Liao, C. P. Lee, Y. H. Chiu, and Z. H. Wei, "Influence of different onion states on magnetization reversal processes in permalloy rings," *IEEE Transactions on Magnetics*, vol. 46, pp. 179–182, Feb 2010.

[13] Z. H. Wei, C. K. Lo, D. R. Liu, Y. P. Hsieh, Y. R. Lee, H. R. Shiao, Y. H. Chiu, C. P. Lee, C. N. Liao, **Y.-J. Chen**, C. J. Hsu, and M. F. Lai, "Hysteresis in a microactuator with single-domain magnetic thin films," *IEEE Transactions on Magnetics*, vol. 46, pp. 630–633, Feb 2010.

[14] Z.-H. Wei, Y.-P. Hsieh, Y.-R. Lee, C.-D. Lin, C.-P. Lee, C.-N. Liao, **Y.-J. Chen**, and C.-J. Hsu, "Planar actuation hysteresis and magnetic switching in single-domain cantilever beam," *Japanese Journal of Applied Physics*, vol. 48, no. 8R, p. 080209, 2009.

[15] M. F. Lai, Z. H. Wei, **Y.-J. Chen**, C. N. Liao, T. F. Ho, C. H. San, C. P. Lee, and Y. P. Hsieh, "Field evolution of magnetization states with different combinations of vortex cores and anti-vortex cores in different helicities," *IEEE Transactions on Magnetics*, vol. 44, pp. 2715–2717, Nov 2008.

Erklärung an Eides statt

Hiermit erkläre ich, dass ich die vorliegende Dissertation

Confined magnon modes and anisotropic exchange interaction in ultrathin Co films

selbstädig und ohne fremde Hilfe verfasst und keine anderen als die von mir angegebenen Quellen und Hilfsmittel benutzt zu haben. Die den benutzten Werken wörtlich oder inhaltlich entnommenen Stellen sind als solche kenntlich gemacht.

Eine Anmeldung der Promotionsabsicht habe ich an keiner anderen Fakultät einer Universität oder Hochschule beantragt.

Halle (Saale), den 17 April, 2016

Ying-Jiun Chen

Curriculum vitae

■■■■■■■■■■■■■■■■■■■ Personal

Name	Ying-Jiun Chen
Date of Birth	October 31, 1984
Nationality	Taiwan

■■■■■■■■■■■■■■■■■■■ Education

Since 07. 2012 **Ph. D. Candidate**, *Max Planck Institute of Microstructure Physics and Martin-Luther-Universität Halle-Wittenberg.*
Supervisor: Prof. Dr. Jürgen Kirschner

05. 2010 – 06. 2012 **Research Assistant**, *National Synchrotron Radiation Research Center*, Hsinchu, Taiwan.

Research Project:

Study of Spin, Charge and Orbital States in Self-Assembled Perovskite-Spinel Heteroepitaxial Nanostructures Using X-Ray Absorption Spectroscopy

09. 2007 – 04. 2010 **Master in Department of Power Mechanical Engineering**, *National Tsing-Hua University*, Hsinchu, Taiwan.

Master Thesis:

Effect of Nano-scaled Magnetic Structure on Thermal Conductivity

09. 2003 – 06. 2007 **Bachelor in Department of Physics**,
National Changhua University of Education, Changhua, Taiwan.

Bachelor Project:

Synthesis and Optical Properties of Gold Nanorods

Halle (Saale), 17 April, 2016
Ying-Jiun Chen